Praise f
THE EGO TU

"Metzinger's intended audience is the lay reader, and he does a superb job of presenting his theory and introducing philosophical issues related to consciousness." —*Library Journal*

"Thomas Metzinger's *The Ego Tunnel* is a major contribution to this project [of scientific enlightenment]." —*Naturalism.Org*

"Metzinger is crisp in his arguments and has a keen appreciation of essential ideas." —*Bookforum*

"Stunningly original exploration of human consciousness. . . . [*The Ego Tunnel*] really fills the need—given the recent upsurge in theoretical and empirical interest in consciousness—for an account of the subjective or phenomenal dimension of consciousness that is accessible to researchers and students from a variety of disciplines." —*Metapsychology.com*

"Thomas Metzinger is one of the most consistently interesting and best informed philosophers working on consciousness today." —NED BLOCK, Silver Professor, Departments of Philosophy, Psychology, and the Center for Neural Science, New York University

"This brilliant book is a must-read for everybody interested in neuroscience. In *The Ego Tunnel*, Thomas Metzinger shows that philosophical discussions of consciousness can be crystal clear and grounded in fascinating psychological experiments and observations on the impact of brain disorders. He proves that philosophy can be relevant to scientific questions about the mind and brain. Read this book and be inspired."

"Thomas Metzinger, one of the leading philosophers of his generation, has focused his efforts on the problem of consciousness. That is hardly surprising for a philosopher, but his approach is. It differs from that of most colleagues by a fearless embrace of neuroscience. Metzinger's perspective is refreshing and thought provoking."

"*The Ego Tunnel* is a highly readable and enjoyable book about a deep problem: Who are we? A fascinating journey into consciousness that combines creativity and scholarship, it explores our private mental life within the realm of illusions, lucid dreaming, and out of body experiences."

The
Ego Tunnel

THE SCIENCE OF THE MIND
AND THE MYTH OF THE SELF

THOMAS METZINGER

BASIC
BOOKS

A MEMBER OF THE
PERSEUS BOOKS GROUP
New York

Books published by Basic Books are available at special
discounts for bulk purchases in the United States by
corporations, institutions, and other organizations. For more
information, please contact the Special Markets Department
at the Perseus Books Group, 2300 Chestnut Street, Suite 200,
Philadelphia, PA 19103, or call (800) 810-4145, ext. 5000, or
e-mail special.markets@perseusbooks.com.

Set in Warnock Pro by the Perseus Books Group

Library of Congress Cataloging-in-Publication Data

Metzinger, Thomas, 1958-
  The ego tunnel : the science of the mind and the myth of
the self / Thomas Metzinger.
    p. cm.
  Includes bibliographical references and index.
  ISBN 978-0-465-04567-9 (alk. paper)
  1. Consciousness. I. Title.

  BF311.M47 2009
  126—dc22
                                    2008044399
  ISBN 978-0-465-02069-0

10 9 8 7 6 5 4 3 2 1

To Anja and my family

Any theory that makes progress is bound to be *initially* counterintuitive.

—DANIEL C. DENNETT, *The Intentional Stance* (Cambridge, MA 1987, p. 6)

He [Ludwig Wittgenstein] once greeted me with the question: "Why do people say that it was natural to think that the sun went round the earth rather than that the earth turned on its axis?" I replied: "I suppose, because it looked as if the sun went round the earth." "Well," he asked, "what would it have looked like if it had *looked* as if the earth turned on its axis?"

—ELIZABETH ANSCOMBE, *An Introduction to Wittgenstein's Tractatus* (London 1959, p. 151)

CONTENTS

ACKNOWLEDGMENTS

This book has not been written for philosophers or scientists. Instead, it is my first attempt to introduce a wider public to what I think are the truly important issues in consciousness research today. The selection of relevant philosophical issues and new empirical insights is entirely my own—and of course hopelessly incomplete and necessarily superficial. But I do hope this book will give interested lay readers a realistic view of the picture of self-consciousness and the human mind now emerging—and of the accompanying challenges all of us will have to face in the future.

Of the many people who have supported me in this project, my first thanks go to Jennifer Windt, who has spent countless hours helping me with the English version. I have learned a great deal from her. It is difficult to write a nonacademic book in a language other than your own, and if I have at least partly succeeded, it is due to her accuracy, conscientiousness, and reliability. I then found a superbly professional editor in Sara Lippincott. I am deeply indebted to both of them. Among the many colleagues who have supported me, I am particularly grateful to Susan Blackmore, Olaf Blanke, Peter Brugger, Daniel Dennett, Vittorio Gallese, Allan Hobson, Victor Lamme, Bigna Lenggenhager, Antoine Lutz, Angelo Maravita, Wolf Singer, Tej Tadi, and Giulio Tononi. This work was in part supported by the COGITO Foundation (Switzerland), the DISCOS Project ("Disorders and Coherence of the Embodied Self," an EU Marie Curie Research Training Network), and a Fellowship at Europe's best Institute for Advanced Research, the *Wissenschaftskolleg zu Berlin.*

Thomas Metzinger
January 2009

INTRODUCTION

In this book, I will try to convince you that there is no such thing as a self. Contrary to what most people believe, nobody has ever *been* or *had* a self. But it is not just that the modern philosophy of mind and cognitive neuroscience together are about to shatter the myth of the self. It has now become clear that we will never solve the philosophical puzzle of consciousness—that is, how it can arise in the brain, which is a purely physical object—if we don't come to terms with this simple proposition: that to the best of our current knowledge there is no thing, no indivisible entity, that is *us*, neither in the brain nor in some metaphysical realm beyond this world. So when we speak of conscious experience as a *subjective* phenomenon, what is the entity *having* these experiences?

There are other important issues in the quest to probe our inner nature—new, exciting theories about emotions, empathy, dreaming, rationality, recent discoveries about free will and the conscious control of our actions, even about machine consciousness—and they are all valuable, as the building blocks of a deeper understanding of ourselves. I will touch on many of them in this book. What we currently lack, however, is the big picture—a more general framework we can work with. The new mind sciences have generated a flood of relevant data but no model that can, at least in principle, integrate all these data. There is one central question we have to confront head on: Why is there always someone *having* the experience? Who is the feeler of your feelings and the dreamer of your dreams? Who is the agent doing the doing, and what is the entity thinking your thoughts? Why is your conscious reality *your* conscious reality?

This is the heart of the mystery. If we want not just the building blocks but a unified whole, these are the essential questions. There is a new story, a provocative and perhaps shocking one, to be told about this mystery: It is the story of the Ego Tunnel.

The person telling you this story is a philosopher, but one who has closely cooperated with neuroscientists, cognitive scientists, and re-searchers in artificial intelligence for many years. Unlike many of my philosopher colleagues, I think that empirical data are often directly rele-vant to philosophical issues and that a considerable part of academic phi-losophy has ignored such data for much too long. The best philosophers in the field clearly are analytical philosophers, those in the tradition of Gottlob Frege and Ludwig Wittgenstein: In the past fifty years, the strongest contributions have come from analytical philosophers of mind. However, a second aspect has been neglected too much: *phenomenology,* the fine-grained and careful description of inner experience as such. In particular, altered states of consciousness (such as meditation, lucid dreaming, or out-of-body experiences) and psychiatric syndromes (such as schizophrenia or Cotard's syndrome, in which patients may actually believe they do not exist) should not be philosophical taboo zones. Quite the contrary: If we pay more attention to the wealth and the depth of conscious experience, if we are not afraid to take consciousness seriously in all of its subtle variations and borderline cases, then we may discover exactly those conceptual insights we need for the big picture.

In the chapters that follow, I will lead you through the ongoing Con-sciousness Revolution. Chapters 1 and 2 introduce basic ideas of con-sciousness research and the inner landscape of the Ego Tunnel. Chapter 3 examines out-of-body experiences, virtual bodies, and phantom limbs. Chapter 4 deals with ownership, agency, and free will; chapter 5 with dreams and lucid dreaming; chapter 6 with empathy and mirror neu-rons; and chapter 7 with artificial consciousness and the possibility of postbiotic Ego Machines. All these considerations will help us to further map out the Ego Tunnel. The two final chapters address some of the consequences of these new scientific insights into the nature of the conscious mind-brain: the ethical challenges they pose and the social and cultural changes they may produce (and sooner than we think), given the naturalistic turn in the image of humankind. I close by arguing that ultimately we will need a new "ethics of consciousness." If we arrive at a comprehensive theory of consciousness, and if we develop ever more sophisticated tools to alter the contents of subjective experience, we will have to think hard about what a *good* state of consciousness is. We urgently need fresh and convincing answers to questions like the following: Which states of consciousness do we want our children to have? Which states of consciousness do we want to foster, and which do we want to ban on ethical grounds? Which states of consciousness can we inflict upon animals, or upon machines? Obviously, I cannot provide definitive answers to such questions; instead, the concluding chapters are meant to draw attention to the important new discipline of neuroethics while at the same time widening our perspective.

THE PHENOMENAL SELF-MODEL

Before I introduce the Ego Tunnel, the central metaphor that will guide the discussion from here onward, it will be helpful to consider an experiment that strongly suggests the purely experiential nature of the self. In 1998, University of Pittsburgh psychiatrists Matthew Botvinick and Jonathan Cohen conducted a now-classic experiment in which healthy subjects experienced an artificial limb as part of their own body.[1] The subjects observed a rubber hand lying on the desk in front of them, with their own corresponding hand concealed from their view by a screen. The visible rubber hand and the subject's unseen hand were then synchronously stroked with a probe. The experiment is easy to replicate: After a certain time (sixty to ninety seconds, in my case), the famous rubber-hand illusion emerges. Suddenly, you experience the rubber hand as your own, and you feel the repeated strokes in this rubber hand. Moreover, you feel a full-blown "virtual arm"—that is, a connection from your shoulder to the fake hand on the table in front of you.

The most interesting feature I noticed when I underwent this experiment was the strange tingling sensation in my shoulder shortly before

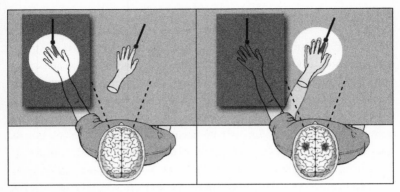

Figure 1: The rubber-hand illusion. A healthy subject experiences an artificial limb as part of her own body. The subject observes a facsimile of a human hand while her own hand is concealed (gray square). Both the artificial rubber hand and the invisible hand are stroked repeatedly and synchronously with a probe. The light areas around the hand and the dark areas in the index finger indicate the respective tactile and visual receptive fields for neurons in the premotor cortex. The illustration on the right shows the subject's illusion as the felt strokes are aligned with the seen strokes of the probe (the dark areas show areas of heightened activity in the brain; the phenomenally experienced, illusory position of the arm is indicated by the light outline). The resulting activation of neurons in the premotor cortex is demonstrated by experimental data. (M. Botvinick & J. Cohen, "Rubber Hand 'Feels' Touch That Eyes See," *Nature* 391:756, 1998.) Figure by Litwak illustrations studio, 2004.

the onset of the illusion—shortly before, as it were, my "soul arm" or "astral limb" slipped from the invisible physical arm into the rubber hand. Of course, there is no such thing as a ghostly arm, and probably no such thing as an astral body, either. What you feel in the rubber-hand illusion is what I call the content of the *phenomenal self-model* (PSM)—the conscious model of the organism as a whole that is activated by the brain. ("Phenomenal" is used here, and throughout, in the philosophical sense, as pertaining to what is known purely experientially, through the way in which things subjectively *appear* to you.) The content of the PSM is the Ego.

The PSM of *Homo sapiens* is probably one of nature's best inventions. It is an efficient way to allow a biological organism to consciously conceive of itself (and others) as a whole. Thus it enables the organism to interact with its internal world as well as with the external environment in an intelligent and holistic manner. Most animals are conscious to one degree or another, but their PSM is not the same as ours. Our evolved type of conscious self-model is unique to the human brain, in that by representing the process of representation itself, we can catch ourselves—as Antonio Damasio would call it—in the act of knowing. We mentally represent ourselves *as* representational systems, in phenomenological real-time. This ability turned us into thinkers of thoughts and readers of minds, and it allowed biological evolution to explode into cultural evolution. The Ego is an extremely useful instrument—one that has helped us understand one another through empathy and mindreading. Finally, by allowing us to externalize our minds through cooperation and culture, the Ego has enabled us to form complex societies.

What lessons can be learned from the rubber-hand illusion? The first point is simple to understand: Whatever is part of your PSM, whatever is part of your conscious Ego, is endowed with a feeling of "mineness," a conscious sense of ownership. It is experienced as *your* limb, *your* tactile sensation, *your* feeling, *your* body, or *your* thought. But then there is a deeper question: Isn't there something more to the conscious self than the mere subjective experience of ownership for body parts or mental states? Isn't there something like "global ownership," a deeper sense of selfhood having to do with owning and controlling your body *as a whole?* What about the experience of *identifying* with it? Could this deep sense of selfhood perhaps be experimentally manipulated? When I first experienced the rubber-hand illusion, I immediately thought it would be important to see whether this would also work with a whole rubber body or an image of yourself. Could one create a full-body analog of the rubber-hand illusion? Could the entire self be transposed to a location outside of the body?

As a matter of fact, there are phenomenal states in which people have the robust feeling of being outside their physical body—these are the so-called out-of-body experiences, or OBEs. OBEs are a well-known class of states in which one undergoes the highly realistic illusion of leaving one's physical body, usually in the form of an etheric double, and moving outside of it. Phenomenologically, the subject of experience is located in this double. Obviously, if one seriously wants to understand what the conscious self is, these experiences are of great philosophical and scientific relevance. Could they be created in the lab?

One of the neuroscientists I am proud to collaborate with is Olaf Blanke, a brilliant young neurologist at the Swiss Federal Institute of Technology in Lausannne, who was the first scientist to trigger an OBE by directly stimulating the brain of a patient with an electrode. There are typically two representations of one's body in these experiences: the visual one (the sight of your own body, lying on the bed, say, or on an operating table) and the felt one, in which you feel yourself to be hovering above or floating in space. Interestingly, this second body-model is the content of the PSM. This is where the Ego is. In a series of virtual-reality experiments, Olaf, his PhD student Bigna Lenggenhager, and I attempted to create artificial OBEs and full-body illusions (see chapter 3).[2] During these illusions, subjects localized themselves outside their body and transiently identified with a computer-generated, external image of it. What these experiments demonstrate is that the deeper, holistic sense of self is not a mystery immune to scientific exploration—it is a form of conscious representational content, and it can be selectively manipulated under carefully controlled experimental conditions.

Throughout the book, I use one central metaphor for conscious experience: the "Ego Tunnel." Conscious experience is like a tunnel. Modern neuroscience has demonstrated that the content of our conscious experience is not only an internal construct but also an extremely selective way of representing information. This is why it is a tunnel: What we see and hear, or what we feel and smell and taste, is only a small fraction of what actually exists out there. Our conscious model of reality is a low-dimensional projection of the inconceivably richer physical reality surrounding and sustaining us. Our sensory organs are limited: They evolved for reasons of survival, not for depicting the enormous wealth and richness of reality in all its unfathomable depth. Therefore, the ongoing process of conscious experience is not so much an image of reality as a tunnel *through* reality.

Whenever our brains successfully pursue the ingenious strategy of creating a unified and dynamic inner portrait of reality, we become conscious. First, our brains generate a world-simulation, so perfect that we do not recognize it as an image in our minds. Then, they generate an inner image of ourselves as a whole. This image includes not only our body and our psychological states but also our relationship to the past and the future, as well as to other conscious beings. The internal image of the person-as-a-whole is the phenomenal Ego, the "I" or "self" as it appears in conscious experience; therefore, I use the terms "phenomenal Ego" and "phenomenal self" interchangeably. The phenomenal Ego is not some mysterious thing or little man inside the head but the content of an inner image—namely, the conscious self-model, or PSM. By placing the self-model within the world-model, a center is created. That center is what we experience as ourselves, the Ego. It is the origin of what philosophers often call the first-person perspective. We are not in direct contact with outside reality or with ourselves, but we do have an inner perspective. We can use the word "I." We live our conscious lives in the Ego Tunnel.

In ordinary states of consciousness, there is always someone *having* the experience—someone consciously experiencing himself as directed toward the world, as a self in the act of attending, knowing, desiring, willing, and acting. There are two major reasons for this. First, we possess an integrated inner image of ourselves that is firmly anchored in our feelings and bodily sensations; the world-simulation created by our brains includes the experience of a *point of view*. Second, we are unable to experience and introspectively recognize our self-models *as* models; much of the self-model is, as philosophers might say, *transparent*.[3] Transparency simply means that we are unaware of the medium through which information reaches us. We do not see the window but only the bird flying by. We do not see neurons firing away in our brain but only what they represent for us. A conscious world-model active in the brain is transparent if the brain has no chance of discovering that it is a model—we look right through it, directly onto the world, as it were. The central claim of this book—and the theory behind it, the *self-model theory of subjectivity*[4]—is that the conscious experience of being a self emerges because a large part of the PSM in your brain is transparent.

The Ego, as noted, is simply the content of your PSM at this moment (your bodily sensations, your emotional state, your perceptions, memories, acts of will, thoughts). But it can become the Ego only because you are constitutionally unable to realize that all this is just the content of a simulation in your brain. It is not reality itself but an image of reality—and a very special one indeed. The Ego is a transparent mental image: You—the physical person as a whole—look right through it. You do not *see* it. But you see *with* it. The Ego is a tool for controlling and planning your behavior and for understanding the behavior of others. Whenever the organism needs this tool, the brain activates a PSM. If—as, for instance, in dreamless deep sleep—the tool is not needed anymore, it is turned off.

It must be emphasized that although our brains create the Ego Tunnel, no one lives in this tunnel. We live with it and through it, but there is no little man running things inside our head. The Ego and the Tunnel are evolved representational phenomena, a result of dynamical self-organization on many levels. Ultimately, subjective experience is a biological data format, a highly specific mode of presenting information about the world by letting it appear as if it were an Ego's knowledge. But no such things as selves exist in the world. A biological organism, as such, is not a self. An Ego is not a self, either, but merely a form of representational content—namely, the content of a transparent self-model activated in the organism's brain.

Variations of this tunnel metaphor illustrate other new ideas in mind science: What would it mean for an Ego Tunnel to branch out to include other Ego Tunnels? What happens to the Ego Tunnel during the dream state? Can machines possess an artificial form of self-consciousness, and can they develop a proper Ego Tunnel? How do empathy and social cognition work; how can communication take place from one tunnel to the next? Finally, of course, we must ask: Is it possible to *leave* the Ego Tunnel?

The idea of an Ego Tunnel is based on an older notion that has been around for quite some time now. It is the concept of a "reality tunnel," which can be found in research on virtual reality and the programming of advanced video games, or in the popular work of nonacademic philosophers such as Robert Anton Wilson and Timothy Leary. The general idea is this: Yes, there is an outside world, and yes, there is an objective reality, but in moving through this world, we constantly apply unconscious filter mechanisms, and in doing so, we unknowingly construct our own individual world, which is our "reality tunnel." We are never directly in touch with reality as such, because these filters prevent us from seeing the world as it is. The filtering mechanisms are our sensory systems and our brain, the architecture of which we inherited from our biological ancestors, as well as our prior beliefs and implicit assumptions. The construction process is largely invisible; in the end, we see only what our reality tunnel allows us to see, and most of us are completely unaware of this fact.

From a philosopher's point of view, there is a lot of nonsense in this popular notion. We don't create an individual world but only a world-model. Moreover, the whole idea of potentially being directly in touch with reality is a sort of romantic folklore; we know the world only by using representations, because (correctly) representing something is what knowing *is*. Also, the Ego Tunnel is not about what psychologists call "confirmation bias"—that is, our tendency to notice and assign significance to observations that confirm our beliefs and expectations, while filtering out or rationalizing away observations that do not. Nor is it true that we can never get out of the tunnel or know anything about the outside world: Knowledge is possible, for instance, through the cooperation and communication of large groups of people—scientific communities that design and test theories, constantly criticize one another, and exchange empirical data and new hypotheses. Finally, the popular notion of a reality tunnel is playfully used in simply too many ways and contexts and therefore remains hopelessly vague.

In the first chapter, I confine discussion to the phenomenon of *conscious experience* and develop a better and richer understanding of why exactly it is exclusively internal. One question to be addressed is, How can all this take place inside the brain and at the same time create the robust experience of living in a reality that is experienced as an external reality? We want to understand how what Finnish philosopher and neuroscientist Antti Revonsuo calls an "out-of-brain experience" is possible:

the experience you have all the time—for instance, right now, as you are reading this book. The robust experience of *not* being in a tunnel, of being directly and immediately in touch with external reality, is one of the most remarkable features of human consciousness. You even have it during an out-of-body experience.

To confine oneself to studying consciousness *as such* means to consider the phenomenal content of one's mental representations—that is, how they feel to you from the first-person perspective, what it is like (subjectively, privately, inwardly) to have them. For example, the predominant phenomenal content of seeing a red rose is the quality of redness itself. In the conscious experience of smelling a mixture of amber and sandalwood, the phenomenal content is that raw subjective quality of "amber-ness" and "sandalwood-ness," ineffable and apparently simple. In experiencing an emotion—say, feeling happy and relaxed—the phenomenal content is the feeling itself and not whatever it refers to.

All evidence now points to the conclusion that phenomenal content is determined locally, not by the environment at all but by internal properties of the brain only. Moreover, the relevant properties are the same regardless of whether the red rose is there in front of you or merely imagined or dreamed about. The subjective sandalwood-and-amber experience doesn't require incense, it doesn't even require a nose; in principle it can also be elicited by stimulating the right combination of glomeruli in your olfactory bulb. Glomeruli (there are some two thousand of them) take input from one type or another of your olfactory receptor cells. If the unified sensory quality of smelling sandalwood and amber typically involves activating smell receptor cells of type 18, 93, 143, and 211 in your nose, then we would expect to get the same conscious experience—an identical odor—by stimulating the corresponding glomeruli with an electrode. The question is, What is the minimally sufficient set of neural properties? Could we selectively elicit exactly the same phenomenon by doing even less, possibly at another location in the brain? Most neuroscientists, and probably the majority of philosophers as well, would answer *yes:* Activate the minimal neural correlate of a given conscious experience and you get the conscious experience itself.

The same general idea holds for more complex states: Their phenomenal content is precisely that aspect of a state (say, of happiness plus relaxation) that not only emerges naturally in everyday situations but can also be caused by a psychoactive substance—or, at least in principle, triggered by an evil neuroscientist experimenting on a living brain in a vat. The problem of consciousness is all about subjective experience, about the structure of our inner life, and not about knowledge of the outer world.

One way of looking at the Ego Tunnel is as a complex property of the global neural correlate of consciousness (NCC). The NCC is that set of neurofunctional properties in your brain sufficient to bring about a conscious experience. There is a specific NCC for the redness of the rose you experience, another for the perceptual object (that is, the rose as a whole), and yet another underlying your accompanying feeling of happiness and relaxation. But there is also a *global* NCC—that is, a much larger set of neural properties underlying consciousness as a whole, underpinning your experiential model of the world, the totality of everything you subjectively feel. The incessant information flow in this global NCC is what creates the tunnel, the world in which you live your conscious life.

But what is this "you"? As I claimed at the outset, we will never have a truly satisfying comprehensive scientific theory of the human mind if we don't dissolve the core of the problem. If we want everything to fall into place—if we want to understand the big picture—then this is the challenge. Why is consciousness *subjective?* The most important question I seek to answer is why a conscious world-model almost invariably has a center: a *me*, an Ego, an experiencing self. What exactly is the self that has the rubber-hand illusion? What exactly is it that apparently leaves the physical body in an OBE? What exactly is it that is reading these lines right now?

An Ego Tunnel is a consciousness tunnel that has evolved the additional property of creating a robust first-person perspective, a subjective view of the world. It is a consciousness tunnel plus an apparent self. This is the challenge: If we want the big picture, we need to know how a genuine sense of selfhood appears. We have to explain your experience of *yourself* as feeling the tactile sensation in the rubber hand, of *yourself* as understanding the sentences you're reading right now. This genuine conscious sense of selfhood is the deepest form of inwardness, much deeper than just being "in the brain" or "in a simulated world in the brain." This nontrivial form of inwardness is what this book is about.

PART ONE | # THE CONSCIOUSNESS PROBLEM

ONE

✦

THE APPEARANCE OF A WORLD

Consciousness is the *appearance of a world*. The essence of the phenomenon of conscious experience is that a single and unified reality becomes present: If you are conscious, a world appears to you. This is true in dreams as well as in the waking state, but in dreamless deep sleep, nothing appears: The fact that there is a reality out there and that you are present in it is unavailable to you; you do not even know that you exist.

Consciousness is a very special phenomenon, because it is part of the world and contains it at the same time. All our data indicate that consciousness is part of the physical universe and is an evolving biological phenomenon. Conscious experience, however, is much more than physics plus biology—more than a fantastically complex, dancing pattern of neural firing in your brain. What sets human consciousness apart from other biologically evolved phenomena is that it makes a reality appear *within itself*. It creates inwardness; the life process has become aware of itself.

Judging from the available data on animal brains and evolutionary continuity, the *appearance of worlds* in biological nervous systems is a recent phenomenon, perhaps only a few million years old. In Darwinian evolution, an early form of consciousness might have arisen some 200 million years ago in the primitive cerebral cortices of mammals, giving them bodily awareness and the sense of a surrounding world and guiding their behavior. My intuition is that birds, reptiles, and fish have long had some sort of awareness too. In any case, an animal that cannot reason or speak a language can certainly have transparent phenomenal states—and that is all it takes to make a world appear in consciousness. Such well-known consciousness researchers and theoretical neurobiologists as Anil Seth, Bernard Baars, and D. B. Edelman have established seventeen criteria for brain structures subserving consciousness, and the evidence for the existence of such structures not only in mammals but also in birds and potentially in octopi is overwhelming. The empirical evidence for animal consciousness is now far beyond any reasonable doubt.[1] Like us, animals are naive realists, and if they have, say, color sensations, it is plausible to assume that these appear to them with the same quality of directness, certainty, and immediacy as they do to us. But the philosophical point is that we really do not know. These are exactly the sort of questions we can consider only after we have constructed a satisfactory theory of consciousness.

A much more recent phenomenon emerged only a couple of thousand years ago—the conscious formation of theories in the minds of human philosophers and scientists. Thus the life process became reflected not only in conscious individual organisms but also in groups of human beings trying to understand the emergence of self-conscious minds as such—that is, what it means that something can "appear within itself." The most fascinating feature of the human mind, perhaps, is not simply that it can sometimes be conscious, or even that it allows for the emergence of a PSM. The truly remarkable fact is that we can also attend to the content of our PSM and form concepts about it. We can communicate about it with one another, and we can experience this as our *own* activity. The process of attending to our thoughts and emotions, to our perceptions and bodily sensations, is itself integrated into the self-model. This property, as noted, probably distinguishes us from most other animals on this planet: the ability to turn the first-person perspective inward, to explore our emotional states and attend to our cognitive processes. As philosophers say, these are "higher-order" levels of the

PSM. They allowed us to become aware of the fact that we are representational systems.

Over the centuries, the theories we have devised have gradually changed our image of ourselves, and in so doing they have subtly altered the contents of consciousness. True, consciousness is a robust phenomenon; it doesn't change simply because of the opinions we have about it. But it does change through practice (think of wine connoisseurs, perfume designers, musical geniuses). Human beings in other historical epochs—during the Vedic period of ancient India, say, or during the European Middle Ages, when God was still perceived as a real and constant presence—likely knew kinds of subjective experience almost inaccessible to us today. Many deep forms of conscious self-experience have become all but impossible due to philosophical enlightenment and the rise of science and technology—at least for the many millions of well-educated, scientifically informed people. Theories change social practice, and practice eventually changes brains, the way we perceive the world. Through the theory of neural networks, we have learned that the distinction between structure and content—between the carrier of a mental state and its meaning—is not as clear-cut as is often assumed. Meaning does change structure, though slowly. And the structure in turn determines our inner lives, the flow of conscious experience.

In the early 1970s, after the heyday of behaviorism, interest in consciousness as a serious research topic began to rise. In several scientific disciplines, the topic of subjective experience gradually became a secret research frontier. Then, in the last decade of the twentieth century, a number of eminent neuroscientists accepted consciousness as a proper target for rigorous research. Now things developed very quickly. In 1994, after a conference of consciousness researchers in Tucson, Arizona, I helped found a new organization, the Association for the Scientific Study of Consciousness (ASSC), which is aimed at drawing together the more rigorous researchers in science and philosophy. The number of conferences and journal articles increased steeply.[2] The following year, I edited a collection of philosophical articles entitled *Conscious Experience*.[3] When one of my ASSC cofounders, Australian philosopher David Chalmers, and I compiled the bibliography, spanning the period

1970–1995, it contained about a thousand entries. Ten years later, when I updated this bibliography for the fifth German edition, it had almost twenty-seven hundred entries. At this point, I gave up my attempt to include all of the new literature on consciousness; it was simply no longer possible. The field is now well established and developing steadily.

In the meantime, we have learned many lessons. We have learned how great the fear of reductionism is, in the humanities as well as among the general public, and how immense the market is for mysterianism. The straightforward philosophical answer to the widespread fear that philosophers or scientists will "reduce consciousness" is that reduction is a relationship between theories, not phenomena. No serious empirical researcher and no philosopher wants to "reduce consciousness"; at best, one theory about how the contents of conscious experience arose can be reduced to another theory. Our theories about phenomena change, but the phenomena stay the same. A beautiful rainbow continues to be a beautiful rainbow even after it has been explained in terms of electromagnetic radiation. Adopting a primitive scientistic ideology would be just as bad as succumbing to mysterianism. Furthermore, most people would agree that the scientific method is not the only way of gaining knowledge.

But this is not the whole story. Frequently, a deeper, unarticulated insight may lie behind our uneasiness with reductive approaches to the conscious mind. We know that our beliefs about consciousness can subtly change what we perceive, influencing the very contents and functional profile of subjective experience itself. Some fear that a materialistic disenchantment, along with advances in the sciences of the mind, may have unwanted social and cultural consequences. As I point out in the concluding chapters of this book, these voices are absolutely right: This is an important aspect of the development of the mind sciences. We have learned that consciousness—like science itself—is a culturally embedded phenomenon.

We have also come to understand that consciousness is not an all-or-nothing affair, a phenomenon that either does or does not exist. It is a graded phenomenon and comes in many different shades. Consciousness is also not a unitary phenomenon but has many discernible aspects: memory, attention, feelings, the perception of color, self-awareness, and higher-order thought. Nevertheless, the essence of the phenomenon— what I call the *appearance of a world*—seems to be preserved throughout. One of the essential features of consciousness is that it situates you in this world. When you wake up in the morning, you experience yourself as existing at a specific time, at a single location, and embedded in a scene: A single and integrated *situation* emerges. The same is true for dreams or hallucinations, in which you not only experience yourself but also experience yourself in the context of a particular situation, as part of a world that has just appeared. We have learned that consciousness reaches down into the animal kingdom.[4] We have learned about psychiatric disorders and brain lesions, about coma and minimally conscious states, about dreams, lucid dreams, and other altered states of consciousness. All this has led to a general picture of a complex phenomenon that comes in different flavors and strengths. There is no single on-off switch. The fact that consciousness is a graded phenomenon sometimes causes conceptual problems. At the same time, we are already beginning to find the first neural correlates of specific forms of conscious content.[5] Eventually we should be able to discern the minimal set of properties our brains require to activate specific qualities of experience, such as the apricot-pink color of the evening sky or the scent of amber and sandalwood.

However, what we do not know is how far discovering such neural correlates will go toward *explaining* consciousness. Correlation is not causation, nor is it explanation. And if certain aspects of consciousness are ineffable, we obviously cannot correlate them with states in our brains. We have no good understanding of what it means to say that consciousness is "subjective," a "private" phenomenon tied to individual selves. But pinning down the neural correlates of specific conscious contents will lay the foundation for future neurotechnology. As soon as we know the sufficient physical correlates of apricot-pink or sandalwood-amber, we will in principle be able to activate these states by stimulating the brain in an appropriate manner. We will be able to modulate our sensations of color or smell, and intensify or extinguish them, by stimulating or inhibiting the relevant groups of neurons. This may also be true for emotional states, such as empathy, gratitude, or religious ecstasy.

First things first, however. Before we can understand what the self is, we must look at the current status of consciousness science by taking a brief tour of the landscape of consciousness, with its unique complex of problems. There has been considerable progress, but as far as our conscious minds are concerned, we still live in prehistoric times. Our theories about consciousness are as naive as the first ideas cavemen probably had about the true nature of the stars. Scientifically, we are at the very beginning of a true science of consciousness.

The conscious brain is a biological machine—a reality engine—that purports to tell us what exists and what doesn't. It is unsettling to discover that there are no colors out there in front of your eyes. The apricot-pink of the setting sun is not a property of the evening sky; it is a property of the internal *model* of the evening sky, a model created by your brain. The evening sky is colorless. The world is not inhabited by colored objects at all. It is just as your physics teacher in high school told you: Out there, in front of your eyes, there is just an ocean of electromagnetic radiation, a wild and raging mixture of different wavelengths. Most of them are invisible to you and can never become part of your conscious model of reality. What is really happening is that the visual system in your brain is drilling a tunnel through this inconceivably rich physical environment and in the process is painting the tunnel walls in various shades of color. *Phenomenal* color. *Appearance.* For your conscious eyes only.

Still, this is only the beginning. There is no clean one-to-one mapping of consciously experienced colors to physical properties "out there." Many different mixtures of wavelengths can cause the same sensation of apricot-pink (scientists call these mixtures *metamers*). It is interesting to note how the perceived colors of objects stay relatively constant under varying conditions of illumination. An apple, for instance, looks green to us at midday, when the main illumination is white sunlight, and also at sunset, when the main illumination is red with a lot of yellow. Subjective color constancy is a fantastic feature of human color perception, a major neurocomputational achievement. On the other hand, you can consciously experience the same physical property, say, the hot kitchen stove in front of you, as two different conscious qualities. You can experience it as the sensation of warmth and as the sensation of glowing red, as something you feel on your skin and as something you project into a space in front of your eyes.

Nor must your eyes be open to enjoy color experience. Obviously, you can also dream of an apricot-pink evening sky, or you can hallucinate one. Or you can enjoy an even more dramatic color experience under the influence of a hallucinogenic drug, while staring into the void behind your closed eyelids. Converging data from modern consciousness research show that what is common to all possible conscious sensations of *apricot-pink* is not so much the existence of an object "out there" as a highly specific pattern of activation in your brain. In principle, you could have this experience without eyes, and you could even have it as a disembodied brain in a vat. What makes you so sure you are not in a vat right now, while you're reading this book? How can you prove that the book in your hand—or your hand itself, for that matter—really exists? (In philosophy, we call this game *epistemology*—the theory of knowledge. We have been playing it for centuries.)

Conscious experience, as such, is an internal affair. Whatever else may or may not be true about consciousness, once all the internal properties of your nervous system are set, all the properties of your conscious experience—its subjective content and the way it *feels* to you—are fully determined. By "internal" I mean not only spatial but also temporal internality—whatever is taking place right now, at this very moment. As soon as certain properties of your brain are fixed, everything you are experiencing at this very moment is also fixed.

Philosophically, this does not yet mean that consciousness can be explained reductively. Indeed, it is not clear what counts as a whole experience: Are experiences discrete, countable entities? However, the *flow* of experience certainly exists, and cognitive neuroscience has shown that the process of conscious experience is just an idiosyncratic path through a physical reality so unimaginably complex and rich in information that it will always be hard to grasp just how reduced our subjective experience is. While we are drinking in all the colors, sounds, and smells—the diverse range of our emotions and sensory perceptions—it's hard to believe that all of this is merely an internal shadow of something inconceivably richer. But it is.

Shadows do not have an independent existence. And the book you are holding right now—that is, the unified sensations of its color, weight, and texture—is just a shadow, a low-dimensional projection of a higher-dimensional object "out there." It is an image, a representation that can be described as a region in your neural state-space. This state-space itself may well have millions of dimensions; nevertheless, the physical reality you navigate with its help has an inconceivably higher number of dimensions.

The shadow metaphor suggests Book VII of Plato's *Republic*. In Plato's beautiful parable, the captives in the cave are chained down at their thighs and necks. They can only look straight ahead; their heads have been shackled in a fixed position since birth. All they have ever seen of themselves and of one another are the shadows cast on the opposite wall of the cave by the fire burning behind them. They believe the shadows to be real objects. The same is true of the shadows cast by the objects carried along above the wall behind their heads. Might we be like the captives, in that objects from some outside world cast shadows on the wall in front of us? Might we be shadows ourselves? Indeed, the philosophical version of our position on reality developed from Plato's myth—except that our version neither denies the reality of the material world nor assumes the existence of eternal forms constituting the true objects of those shadows on the wall of Plato's cave. It does, however, assume that the images appearing in the Ego Tunnel are dynamic projections of something far greater and richer.

But what is the cave, and what are the shadows? *Phenomenal* shadows are low-dimensional projections within the central nervous system of a biological organism. Let us assume that the book you are holding, as consciously experienced by you at this very moment, is a dynamic, low-dimensional shadow of the actual physical object in your actual physical hands, a dancing shadow in your central nervous system. Then we can ask: What is the fire that causes the projection of flickering shadows of consciousness, dancing as activation patterns on the walls of your neural cave? The fire is neural dynamics. The fire is the incessant, self-regulating flow of neural information-processing, constantly perturbed and modulated by sensory input and cognition. The wall is not a two-dimensional surface but the high-dimensional phenomenal state-space of human Technicolor phenomenology.[6] Conscious experiences are full-blown mental models in the representational space opened up by the gigantic neural network in our heads—and because this space is generated by a person possessing a memory and moving forward in time, it is a tunnel. The pivotal question is this: If something like this is taking place all the time, why don't we ever become aware of it?

Antti Revonsuo alluded to the fascinating phenomenon of OBEs when he compared conscious experience to a constant and effortless *out-of-brain experience.*[7] As I have, he invokes the world-simulation model to explain why the sense of presence you are enjoying right now is only an inner, subjective kind of presence. The idea is that the content of consciousness is the content of a simulated world in our brains, and the sense of *being there* is itself a simulation. Our conscious experience of the world is systematically externalized because the brain constantly creates the experience that *I am present in a world outside my brain.* Everything we know about the human brain today indicates that the experience of being outside the brain, and not in a tunnel, is brought about by neural systems buried deep inside the brain. Of course, an external world does exist, and knowledge and action do causally connect us to it—but the conscious experience of knowing, acting, and being connected is an exclusively internal affair.

Any convincing theory of consciousness will have to explain why this does not seem so to us. Therefore, let us embark on a brief tour of the Ego Tunnel, examining some of the most important problems for a philosophically as well as neuroscientifically convincing theory of consciousness. We will discuss six of them in detail: the *One-World Problem,* or the unity of consciousness; the *Now Problem,* or the appearance of a lived moment; the *Reality Problem,* or why you were born as a naive realist; the *Ineffability Problem,* or what we will never be able to talk about; the *Evolution Problem,* or the question of what consciousness was good for; and finally, the *Who Problem,* or the issue of what is the entity that has conscious experience. We are starting with the easiest problem and ending with the hardest. After this, our groundwork will be done.

✦

A TOUR OF THE TUNNEL

THE ONE-WORLD PROBLEM:
THE UNITY OF CONSCIOUSNESS

Once upon a time, I had to write an encyclopedia article on "Conscious-ness." The first thing I did was to photocopy all existing encyclopedia articles on the topic I could find and track down the historical references. I wanted to know whether in the long history of Western philosophy there was a common philosophical insight running like a thread through humanity's perennial endeavor to understand the conscious mind. To my surprise, I found two such essential insights.

The first is that consciousness is a higher-order form of knowledge accompanying thoughts and other mental states. The Latin concept of *conscientia* is the original root from which all later terminologies in English and the Romance languages developed. This in turn is derived from *cum* ("with," "together") and *scire* ("to know"). In classical antiq-uity, as well as in the scholastic philosophy of the Christian Middle Ages, *conscientia* typically referred either to moral conscience or to knowledge shared by certain groups of people—again, most commonly of moral ideas. Interestingly, being truly conscious was connected to moral insight. (Isn't it a beautiful notion that becoming conscious in the true sense could be related to moral conscience? Philosophers would have a new definition of the entity they call a *zombie*—an amoral person, ethically fast asleep but with eyes wide open.)[1]

In any case, many of the classical theories stated that becoming conscious had to do with installing an ideal observer in your mind, an inner witness providing moral guidance as well as a hidden, entirely private knowledge about the contents of your mental states. Consciousness connected your thoughts with your actions by submitting them to the moral judgment of the ideal observer. Whatever we may think about these early theories of consciousness-as-conscience today, they certainly possessed philosophical depth and great beauty: Consciousness was an inner space providing a point of contact between the real human being and the ideal one inside, the only space in which you could be together with God even before death. From the time of René Descartes (1596–1650), however, the philosophical interpretation of *conscientia* simply as higher-order knowledge of mental states began to predominate. It has to do with certainty; in an important sense, consciousness is knowing *that* you know *while* you know.

The second important insight seems to be the notion of *integration:* Consciousness is what binds things together into a comprehensive, simultaneous whole. If we have this whole, then a world appears to us. If the information flow from your sensory organs is unified, you experience the world. If your senses come apart, you lose consciousness. Philosophers like Immanuel Kant or Franz Brentano have theorized about this "unity of consciousness": What exactly is it that, at every single point in time, blends all the different parts of your conscious experience into one single reality? Today it is interesting to note that the first essential insight—knowing that you know something—is mainly discussed in philosophy of mind,[2] whereas the neuroscience of consciousness focuses on the problem of integration: how the features of objects are bound together. The latter phenomenon—the One-World Problem of dynamic, global integration—is what we must examine if we want to understand the unity of consciousness. But in the process we may discover how both these essential questions—the top-down version discussed in philosophy of mind and the bottom-up version discussed in the neurosciences—are two sides of the same coin.

What would it be like to have the experience of living in many worlds at the same time, of genuine parallel realities opening up in your mind? Would there be parallel observers, too? The One-World Problem is so simple that it is easily overlooked: In order for a world to appear to us, it has to be *one* world first. For most of us, it seems obvious that we live our conscious lives in a single reality, and the world we wake up to every morning is the same world we woke up to the day before. Our tunnel is *one* tunnel; there are no back alleys, side streets, or alternative routes. Only people who have suffered severe psychiatric disorders or have experimented with major doses of hallucinogens can perhaps conceive of what it means to live in more than one tunnel at a time. The unity of consciousness is one of the major achievements of the brain: It is the not-so-simple phenomenological fact that all the contents of your current experience are seamlessly correlated, forming a coherent whole, the world in which you live your life.

But the problem of integration has to be solved on several subglobal levels first. Imagine you are no longer able to bind the various features of a seen object—its color, surface texture, edges, and so on—into a single entity. In a disorder known as *apperceptive agnosia,* no coherent visual model emerges on the level of conscious experience, despite the fact that all the patient's low-level visual processes are intact. Sufferers typically have a fully intact visual field that is consciously perceived, but they are unable to recognize what it is they are looking at. They cannot distinguish shapes from or match shapes with each other, for example, or copy drawings. Apperceptive agnosia is usually caused by a lack of oxygen supply to the brain—for instance, through carbon monoxide poisoning. Patients may well have a coherent, integrated visual world-model, but certain types of visual information are no longer available to them to act upon. On a functional level, they cannot use gestalt grouping cues or figure/ground cues to organize their visual field.[3] Now imagine you are no longer able to integrate your perception of an object with the categorical knowledge that would allow you to identify it, and you consequently cannot subjectively experience *what it is* you are perceiving—as in *asterognosia* (the inability to recognize objects by touch, typically associated with lesions in two regions of the primary somatosensory cortex) or *autotopagnosia* (the inability to identify and name one's own body parts, also associated with cortical lesions). There are also patients suffering from what has been called *disjunctive agnosia*, who cannot integrate seeing and hearing—whose conscious life seems to be taking place in a movie with the wrong soundtrack. As one patient described his experience, someone "was standing in front of me and I could see his mouth moving, but I noticed that the mouth moving did not belong to what I heard."[4]

Now, what if *everything* came apart? There are neurological patients with wounded brains who describe "shattered worlds," but in these cases there is at least some kind of world left—something that could be experienced as having been shattered in the first place. If the unified, multimodal scene—the Here and Now, the situation as such—dissolves completely, we simply go blank. The world no longer appears to us.

A number of new ideas and hypotheses in the neurosciences suggest how this "world-binding" function works. One such is the *dynamical core hypothesis*,[5] which posits that a highly integrated and internally differentiated neurodynamic pattern emerges from the constant background chatter of millions of neurons incessantly firing away. Giulio Tononi, a neuroscientist at the University of Wisconsin–Madison who is a leading advocate of this hypothesis, speaks of a "functional cluster" of neurons, whereas I have coined the concept of *causal density*.[6]

The basic idea is simple: The global neural correlate of consciousness is like an island emerging from the sea—as noted, it is a large set of neural properties underlying consciousness as a whole, underpinning your experiential model of the world in its totality at any given moment. The global NCC has many different levels of description: Dynamically, we can describe it as a coherent island, made of densely coupled relations of cause and effect, emerging from the waters of a much less coherent flow of neural activity. Or we could adopt a neurocomputational perspective and look at the global NCC as something that results from information-processing in the brain and hence functions as a carrier of information. At this point, it becomes something more abstract, which we might envision as an information cloud hovering above a neurobiological substrate. The "border" of this information cloud is functional, not physical; the cloud is physically realized by widely distributed firing neurons in your head. Just like a real cloud, which is made of tiny water droplets suspended in the air, the neuronal activation pattern underlying the totality of your conscious experience is made of millions of tiny electrical discharges and chemical transitions at the synapses. In strict terms, it has no fixed location in the brain, though it is coherent.

But why is it coherent? What holds all the droplets—all the micro-events—together? We do not yet know, but there are some indications that the unified whole appears by virtue of the temporal fine-structure characterizing the conscious brain's activity—that is, the rhythmic dance of neuronal discharges and synchronous oscillations. This is why the border of this whole is a functional border, outlining the island of consciousness in an ocean made up of a myriad of less integrated and less densely coupled neural micro-events. Whatever information is within this cloud of firing neurons is *conscious* information. Whatever is within the cloud's boundary (the "dynamical core") is part of our inner world; whatever is outside of it is not part of our subjective reality. Conscious experience can thus be seen as a special global property of the overall neural dynamics of your brain, a special form of information-processing based on a globally integrated data format.

We also possess the first mathematical instruments that allow us to describe the causal complexity within the dynamical core of consciousness. Technical details aside, they show us how self-organization in our brains strikes an optimal balance between integration and segregation, creating the wonderful richness and diversity of conscious contents and the unity of consciousness at the same time.

What does all this mean? What we want for consciousness is not a uniform state of global synchrony, a state in which many nerve cells simply fire together simultaneously. We find such uniformity in states of unconsciousness such as deep sleep and during epileptic seizures; in these cases, the synchrony wipes out all the internal complexity: It is as if the synchrony had glossed over all the colors and shapes, the objects making up our world. We want large-scale coherence spanning many areas of the brain and flexibly binding many different contents into a conscious hierarchy: the letters into the page, the page into the book, the hand holding the book into your bodily self, and the self sitting in a chair in the room and understanding the words. We want a unity of consciousness that—internally—is as differentiated as possible. On the other hand, maximal differentiation is not optimal, either, because then our world would fall apart into unconnected pieces of mental content and we would lose consciousness. The trick with consciousness is to achieve just the right trade-off between the parts and the whole—and at any single moment a widely distributed network of neurons in the brain seems to achieve just that, as a cloud of single nerve cells, dispersed in space, fire away in intricate patterns of synchronous activity, perhaps with one pattern becoming embedded in the next. Just like the water droplets that form a real cloud, some elements leave the aggregate at any given moment, while others join it. Consciousness is a large-scale, unified phenomenon emerging from a myriad of physical micro-events. As long as a sufficiently high degree of internal correlation and causal coupling allows this island of dancing micro-events in your brain to emerge, you live in a single reality. A single, unified world appears to you.

This emergence can happen during "offline states" as well: In dreams, however, the binding of contents does not work quite as well, which is why your dream reality is frequently so bizarre, why you have difficulty focusing your attention, why scenes follow each other so quickly. Nevertheless, there is still an overall *situation,* you are still present, and that is why phenomenal experience continues. But when you move into deep sleep and the island dissolves back into the sea, your world disappears as well. We humans have known this since Greek antiquity: Sleep is the little brother of death; it means letting go of the world.[7]

One of the intriguing characteristics of current research into consciousness is how old philosophical ideas reappear in the best of cutting-edge neuroscience—in new disguise, as it were. Aristotle and Franz Brentano alike pointed out that consciously perceiving must also mean being aware of the fact that one is consciously perceiving, right now, at this very moment. In a certain sense, we must perceive the perceiving while it happens. If this idea is true, the brain state creating your conscious perception of the book in your hand right now must have two logical parts: one portraying the book and one continuously representing the state itself. One part points at the world, and one at itself. Conscious states could be exactly those states that "metarepresent" themselves while representing something else. This classical idea has logical problems, but the insight itself can perhaps be preserved in an empirically plausible framework.

Work being done by Dutch neuroscientist Victor Lamme in Amsterdam and in Stanislas Dehaene's lab at the NeuroSpin Center in the CEA campus of Saclay and at the Pitié-Salpêtrière Hospital in Paris converges on the central importance of so-called recurrent connections as a functional basis for consciousness.[8] In conscious visual processing, for example, high-level information is dynamically mapped back to low-level information, but it all refers to the same retinal image. Each time your eyes land on a scene (remember, your eye makes about three saccades per second), there is a feedforward-feedback cycle about the current image, and that cycle gives you the detailed conscious percept of that scene. You continuously make conscious snapshots of the world via these feedforward-feedback cycles. In a more general sense, the principle is that the almost continuous feedback-loops from higher to lower areas create an ongoing cycle, a circular nested flow of information, in which what happened a few milliseconds ago is dynamically mapped back to what is coming in right now. In this way, the immediate past continuously creates a context for the present—it filters what can be experienced right now. We see how an old philosophical idea is refined and spelled out by modern neuroscience on the nuts-and-bolts level. A standing context-loop is created. And this may be a deeper insight into the essence of the world-creating function of conscious experience: Conscious information seems to be integrated and unified precisely because the underlying physical process is mapped back onto itself and becomes its own context. If we apply this idea not to single representations, such as the visual experience of an apple in your hand, but to the brain's unified portrait of the world *as a whole,* then the dynamic flow of conscious experience appears as the result of a continuous large-scale application of the brain's prior knowledge to the current situation. If you are conscious, the overall process of perceiving, learning, and living creates a context for itself—and that is how your reality turns into a *lived* reality.

Another fascinating scientific route into the One-World Problem is increasingly receiving attention. It has long been known that in deep meditation the experience of unity and holistic integration is particularly salient. Thus, if we want to know what consciousness is, why not consult those people who cultivate it in its purest form? Or even better, why not use our modern neuroimaging techniques to look directly into their brains while they maximize the unity and holism of their minds?

Antoine Lutz and his colleagues at the W. M. Keck Laboratory for Functional Brain Imaging and Behavior at the University of Wisconsin studied Tibetan monks who had experienced at least ten thousand hours of meditation. They found that meditators self-induce sustained high-amplitude gamma-band oscillations and global phase-synchrony, visible in EEG recordings made while they are meditating.[9] The high-amplitude gamma activity found in some of these meditators seems to be the strongest reported in the scientific literature. Why is this interesting? As Wolf Singer and his coworkers have shown, gamma-band oscillations, caused by groups of neurons firing away in synchrony about forty times per second, are one of our best current candidates for creating unity and wholeness (although their specific role in this respect is still very much debated). For example, on the level of conscious object-perception, these synchronous oscillations often seem to be what makes an object's various features—the edges, color, and surface texture of, say, an apple—cohere as a single unified percept. Many experiments have shown that synchronous firing may be exactly what differentiates an assembly of neurons that gains access to consciousness from one that also fires away but in an uncoordinated manner and thus does not. Synchrony is a powerful causal force: If a thousand soldiers walk over a bridge together, nothing happens; however, if they march across in lockstep, the bridge may well collapse.

The synchrony of neural responses also plays a decisive role in figure-background segregation—that is, the pop-out effect that lets us perceive an object against a background, allowing a new gestalt to emerge from the perceptual scene. Ulrich Ott is Germany's leading meditation researcher, working at the Bender Institute of Neuroimaging at the Justus-Liebig-Universität in Giessen. He confronted me with an intriguing idea: Could deep meditation be the process, perhaps the only process, in which human beings can sometimes turn the global background *into* the gestalt, the dominating feature of consciousness itself? This assumption would fit in nicely with an intuition held by many, among others Antoine Lutz, namely that the fundamental subject/object structure of experience can be transcended in states of this kind.

Interestingly, this high-amplitude oscillatory activity in the brains of experienced meditators emerges over several dozens of seconds. They can't just switch it on; instead, it begins to unfold only when the meditator manages effortlessly to "step out of the way." The full-blown meditative state emerges only slowly, but this is exactly what the theory predicts: As a gigantic network phenomenon, the level of neural synchronization underlying the unity of consciousness will require more time to develop, because the amount of time required to achieve synchronization is proportional to the size of the neural assembly—in meditation, an orchestrated group of many hundreds of million nerve cells must be formed. The oscillations also correlate with the meditators' verbal reports of the intensity of the meditative experience—that is, oscillations are directly related to reports of intensity. Another interesting finding is that there are significant postmeditative changes to the baseline activity of the brain. Apparently, repeated meditative practice changes the deep structure of consciousness. If meditation is seen as a form of mental training, it turns out that oscillatory synchrony in the gamma range opens just the right time window that would be necessary to promote synaptic change efficiently.

To sum up, it would seem that feature-binding occurs when the widely distributed neurons that represent the reflection of light, the surface properties, and the weight of, say, this book start dancing together, firing at the same time. This rhythmic firing pattern creates a coherent cloud in your brain, a network of neurons representing a single object—the book—for you *at a particular moment*. Holding it all together is coherence *in time*. Binding is achieved in the temporal dimension. The unity of consciousness is thus seen to be a dynamic property of the human brain. It spans many levels of organization, it self-organizes over time, and it constantly seeks an optimal balance between the parts and the whole as they gradually unfold. It shows up on the EEG as a slowly evolving global property, and, as demonstrated by our meditators, it can be cultivated and explored from the inside, from the first-person perspective. Please also see the interview with Wolf Singer at the end of this chapter.

But the next problem in formulating a complete theory of consciousness is more difficult.

THE NOW PROBLEM: A LIVED MOMENT EMERGES

Here is something that, as a philosopher, I have always found both fascinating and deeply puzzling: A complete scientific description of the physical universe would not contain the information as to what time is "now." Indeed, such a description would be free of what philosophers call "indexical terms." There would be no pointers or little red arrows to tell you "You are here!" or "Right now!" In real life, this is the job of the conscious brain: It constantly tells the organism harboring it what place is *here* and what time is *now*. This experiential Now is the second big problem for a modern theory of consciousness.[10]

The biological consciousness tunnel is not a tunnel only in the simple sense of being an internal model of reality in your brain. It is also a time tunnel—or, more precisely, a tunnel of presence. Here we encounter a subtler form of inwardness—namely, an inwardness in the temporal domain, subjectively experienced.

The empirical story will have to deal with short-term memory and working memory, with recurrent loops in neural networks, and with the binding of single events into larger temporal gestalts (often simply called the *psychological moment*). The truly vexing aspect of the Now Problem is conceptual: It is very hard to say what exactly the puzzle consists of. At this point, philosophers and scientists alike typically quote a passage from the fourteenth chapter of the eleventh book of St. Augustine's *Confessions*. Here the Bishop of Hippo famously notes, "What then is time? If no one asks me, I know. If I wish to explain it to one that asketh, I know not." The primary difficulty with the Now Problem is not the neuroscience but how to state it properly. Let me try: Consciousness is inwardness in time. It makes the world present for you by creating a new space in your mind—the space of temporal internality. Everything is *in the Now*. Whatever you experience, you experience it as happening *at this moment.*

You may disagree at first: Is it not true that my conscious, episodic memory of my last walk on the beach refers to something in the past? And is it not true that my conscious thoughts and plans about next weekend's trip to the mountains refer to the future? Yes, this is true—but they are always embedded in a conscious model of the self as remembering the starfish on the beach *right now,* as planning a new route to the peak *at this very moment.*

A major function of conscious experience consists, as the great British psychologist Richard Gregory has put it, in "flagging the dangerous present."[11] One essential function of consciousness is to help an organism stay in touch with the immediate present—with all those properties in both itself and the environment that may change fast and unpredictably. This idea relates to a classic concept introduced by Bernard Baars of the Neurosciences Institute in San Diego, best known for his book *A Cognitive Theory of Consciousness,* in which he outlines his global-workspace theory as a model for consciousness. His fruitful metaphor of consciousness as the content of a global workspace of the mind implies that only the critical aspects are represented in consciousness. Conscious information is exactly that information that must be made available for every single one of your cognitive capacities at the same time. You require a conscious representation only if you do not know exactly what will happen next and which capacities (attention, cognition, memory, motor control) you will need to react properly to the challenge around the corner. This critical information must remain active so that different modules or brain mechanisms can access it simultaneously.

My idea is that this simultaneity is precisely why we need the conscious Now. In order to effect this, our brains learned to simulate temporal internality. In order to create a common platform—a blackboard on which messages to our various specialized brain areas can be posted—we need a common frame of reference, and this frame of reference is a temporal one. Although, strictly speaking, no such thing as Now exists in the outside world, it proved adaptive to organize the inner model of the world around such a Now—creating a common temporal frame of reference for all the mechanisms in the brain so that they can access the same information at the same time. A certain point in time had to be represented in a privileged manner in order to be flagged as reality. The past is outside-time, as is the future. But there is also inside-time, *this* time, the Now, the moment you're currently living. All your conscious thoughts and feelings take place in this lived moment.

How are we going to find this special form of inwardness in the biological brain? Of course, conscious time experience has other elements. We experience simultaneity. (And have you ever noticed that you cannot will two different actions at the same moment or simultaneously make two decisions?) We experience succession: of the notes in a piece of music, of two thoughts drifting by in our minds, one after the other. We experience duration: A musical tone or an emotion may stay constant over time. From all this emerges what the neuroscientist Ernst Pöppel, one of the pioneer researchers in this field, and his colleague Eva Ruhnau, director of the University of Munich's Human Science Center, describe as a temporal gestalt:[12] Musical notes can form a motif—a bound pattern of sounds constituting a whole that you recognize as such from one instant to the next. Similarly, individual thoughts can form more complex conscious experiences, which may be described as unfolding patterns of reasoning.

By the way, there is an upper limit to what you can consciously experience as taking place in a single moment: It is almost impossible to experience a musical motif, a rhythmic piece of poetry, or a complex thought that lasts for more than three seconds as a unified temporal gestalt. When I was studying philosophy in Frankfurt, professors typically did not extemporize during their lectures; instead, they read from a manuscript for ninety minutes, firing rounds of excessively long, nested sentences, one after another, at their students. I suspected that these lectures were not aimed at successful communication at all (although they were frequently *about* it) but that this was a kind of intellectual machismo. ("I am going to demonstrate the inferiority of your intelligence to you by spouting fantastically complex and seemingly endless sentences. They will make your short-term buffer collapse, because you cannot integrate them into a single temporal gestalt anymore. You won't understand a thing, and you will have to admit that your tunnel is smaller than mine!")

I assume many of my readers have encountered this type of behavior themselves. It is a psychological strategy we inherited from our primate ancestors, a slightly more subtle form of ostentatious display behavior that made its way into academia. What enables this new kind of machismo is the limited capacity of the moving window of the Now. Looking through this window, we see enduring objects and meaningful chains of events. Underlying all these experiences of duration, succession, and the formation of temporal wholes is the rock-solid bed of *presence*. In order to understand what the *appearance of a world* is, we urgently need a theory of how the human brain generates this temporal sense of presence.

Presence is a necessary condition for conscious experience. If the brain could solve the One-World Problem but not the Now Problem, a world could not appear to you. In a deep sense, appearance *is* simply presence, and the subjective sense of temporal immediacy *is* the definition of an internal space of time.

Is it possible to transcend this subjective Now-ness, to escape the tunnel of presence? Imagine you are lost in a daydream. Completely. Your conscious mind is not "flagging the dangerous present" anymore. Those animals in the history of our planet that did this too often did not stand a chance of becoming our ancestors; they were eaten by other, less pensive animals. But what actually happens at the moment you fully lose contact with your present surroundings, say, in a manifest daydream? You are suddenly somewhere else. Another lived Now emerges in your mind. Now-ness is an essential feature of consciousness.

And, of course, it is an illusion. As modern-day neuroscience tells us, we are never in touch with the present, because neural information-processing itself takes time. Signals take time to travel from your sensory organs along the multiple neuronal pathways in your body to your brain, and they take time to be processed and transformed into objects, scenes, and complex situations. So, strictly speaking, what you are experiencing as the present moment is actually the past.

At this point, it becomes clear why philosophers speak about "phenomenal" consciousness or "phenomenal" experience. A phenomenon is an appearance. The phenomenal Now is the appearance of a Now. Nature optimized our time experience over the last couple of millions of years so that we experience something as taking place *now* because this arrangement is functionally adequate in organizing our behavioral space. But from a more rigorous, philosophical point of view, the temporal inwardness of the conscious Now is an illusion. There *is* no immediate contact with reality.

This point gives us a second fundamental insight into the tunnel-like nature of consciousness: The sense of presence is an internal phenomenon, created by the human brain. Not only are there no colors out there, but there is also no present moment. Physical time flows continuously. The physical universe does not know what William James called the "specious present," nor does it know an expanded, or "smeared," present moment. The brain is an exception: For certain physical organisms, such as us, it has proved viable to represent the path through reality *as if* there were an extended present, a chain of individual moments through which we live our lives. I like James's metaphor, according to which the present is not a knife-edge but a saddleback with a breadth of its own, on which we sit perched and from which we look in two directions into time. Of course, from the illusory smearing of the present moment in human consciousness it does not follow that some kind of nonsmeared present could not exist on the level of physics—but remember, a complete physical description of the universe would not contain the word "now"; there would be no little red arrow telling us "This is your place in the temporal order." The Ego Tunnel is just the opposite of a God's-eye view of the world. It has a Now, a Here—and a Me, *being there now*.

The lived Now has a fascinating double aspect. From an epistemological point of view, it is an illusion (the present is an appearance). The moving window of the conscious Now, though, has proved functionally advantageous for creatures like us: It successfully bundles perception, cognition, and conscious will in a way that selects just the right parameters of interaction with the physical world, in environments like those in which our ancestors fought for survival. In this sense, it is a form of knowledge: functional, nonconceptual knowledge about what will work with this kind of body and these kinds of eyes, ears, and limbs.

What we experience as the present moment embodies implicit knowledge about how we can integrate our sensory perceptions with our motor behavior in a fluid and adaptive manner. However, this type of knowledge applies only to the kind of environment we found on the surface of this planet. Other conscious beings, in other parts of the universe, might have evolved completely different forms of time experience. They might be frozen into an eternal Now or have a fantastically high resolution, living for only a few of our Earth minutes and experiencing more intense individual moments than a million human beings experience in a lifetime. They could be masters of boredom, subjects of an extremely slow passage of time. A good (and more difficult) question is how much room for variation there is in terms of subjective time experience. If my argument is sound, conscious minds can be situated only in one single, real Now at a time—because this is one of the essential features of consciousness. Is it logically possible to live in two or more absolutely equivalent Nows at the same time, to have a subjective perspective originating from *multiple* points in the temporal order? I don't think so, because there would no longer be one single, present "self" who *had* these experiences. Moreover, it's hard to imagine a situation in which experiencing multiple lived presents might have been adaptive. Thus, although no such thing as an extended present exists from a strict philosophical point of view or from the perspective of a physicist, there must be deep biological truths and a profound evolutionary wisdom behind the way conscious beings such as ourselves happen to represent time in the brain.

Even given a radically materialist view of mind and consciousness, one must concede that there is a complex physical property that (as far as we know) exists only in biological nervous systems on this planet. This new property is a virtual window of presence, and it is implemented in the brains of vertebrates and particularly of higher mammals. It is the lived Now. The physical passage of time existed before this property emerged, but then something new was added—a *representation* of time, including an illusory, smeared present, plus the fact that the beings harboring this new property in their brains could not recognize it as a representation. Billions of conscious, time-representing nervous systems created billions of individual perspectives.

At this point, we also touch on a deeper and more general principle running through modern research on consciousness. The more aspects of subjective experience we can explain in a hardheaded, materialistic manner, the more our view of what the self-organizing physical universe itself is will change. Very obviously, and in a strictly no-nonsense, non-metaphorical, and nonmysterious way, the physical universe itself possesses an intrinsic potential for the emergence of subjectivity. Crude versions of objectivism are false, and reality is much richer than we thought.

THE REALITY PROBLEM:
HOW YOU WERE BORN AS A NAIVE REALIST

Minimal consciousness is the appearance of a world. However, if we solve the One-World Problem and the Now Problem, all we have is a *model* of a unified world and a *model* of the present moment in the brain. We have a representation of a single world and a representation of a single moment. Clearly, the appearance of a world is something different. Imagine you could suddenly apprehend the whole world, your own body, the book in your hands, and all of your current surroundings as a "mental model." Would this still be conscious experience?

Now, try to imagine something even more difficult: The robust sense of presence you are enjoying right now is itself only a special kind of image. It is a time representation in your brain—a fiction, not the real thing. What would happen if you could distance yourself from the current moment—if the Now-ness of this current moment turned out not to be the real Now but only an elegant portrait of presence in your mind? Would you still be conscious? This is not simply an empirical issue; it also possesses a distinct philosophical flavor. The pivotal question is how to get from a world-model and a Now-model to exactly what you have as you are reading this: the presence of a world.

The answer lies in the transparency of phenomenal representations. Recall that a representation is transparent if the system using it cannot recognize it as a representation. A world-model active in the brain is transparent if the brain has no chance of discovering that it is a model. A model of the current moment is transparent if the brain has no chance of discovering that it is simply the result of information-processing currently going on in itself. Imagine you are watching a movie on TV—*2001: A Space Odyssey*, say, and you have just watched the scene in which the victorious apeman throws his bone-weapon high into the air, at which point the film jumps into the future, matching the image of the tumbling bone to that of a spacecraft. Dr. Heywood R. Floyd reaches Moon Base Clavius in his lunar landing craft, and discusses with the local Soviet scientists "the potential for culture shock and social disorientation" presented by the discovery of a monolith on the moon. When they arrive at the gigantic black monolith, a member of the exploring party reaches out and strokes its smooth surface, mirroring the awe and curiosity the apemen exhibited millions of years earlier. The scientists and astronauts gather around it for a group photo, but suddenly an earsplitting high-pitched tone is picked up by their earphones—a tone emitted by the monolith as the sun shines down on it. You are completely engaged in the scene unfolding in front of you, to the point of identifying with the bewildered spacesuited humans. However, you can distance yourself from the movie at any time and become aware that there is a separate you sitting on the couch in the living room and only watching all this. You can also move up close to the screen and inspect the little pixels, thousands of little squares of light rapidly blinking on and off, creating a continuous flowing image as soon as you are a couple of yards away. Not only is this flowing image made up of individual pixels, but the temporal dynamic is not really continuous at all—the individual pixels blink on and off according to a certain rhythm, changing their color in abrupt steps.

You cannot do this distancing with your consciousness. It is a different kind of medium. If you look at the book in your hands and try to apprehend individual pixels, you can't see any. The appearance of the book is dense and impenetrable. Visual attention cannot dissolve the fluidity, the continuity, of your book experience as it can discover the individual pixels when you take a closer look at the TV screen. The blinding speed with which your brain activates the visual model of the book and integrates it with the tactile sensations in your fingers is simply too fast.

One might argue that this disparity exists because the system creating the "pixels" is also the one trying to detect them. Of course, in the continuous flow of information-processing in the brain, nothing like pixels really exists. Still, could your inability to break the book percept down into pixels be caused by something other than the speed of integration in the brain? If your brain worked much more slowly (say, if it could detect time spans of a year but no briefer), you still wouldn't be able to detect those "pixels." You would still perceive a seamless passage of time, because the conscious working of our brain is not a single uniform event but a multilayered chain of events in which different processes are densely coupled and interacting all the time. The brain creates what are called *higher-order representations*. If you attend to your perception of a visual object (such as this book), then there is at least one second-order process (i.e., attentional processing) taking a first-order process—in this case, visual perception—as its object. If the first-order process—the process creating the seen object, the book in your hands—integrates its information in a smaller time-window than the second-order process (namely, the attention you're directing at this new inner model), then the integration process on the first-order level will itself become transparent, in the sense that you cannot consciously experience it. By necessity, you are now blind to the fundamental construction process. Transparency is not so much a question of the speed of information-processing as of the speed of different types of processing (such as attention and visual perception) relative to each other.

Just as swiftly and effortlessly, the book-model is bound with other models, such as the models of your hands and of the desk, and seamlessly integrated into your overall conscious space of experience. Because it has been optimized over millions of years, this mechanism is so fast and so reliable that you never notice its existence. It makes your brain invisible to itself. You are in contact only with its content; you never see the representation as such; therefore, you have the illusion of being directly in contact with the world. And that is how you become a *naïve realist,* a person who thinks she is in touch with an observer-independent reality.

If you talk to neuroscientists as a philosopher, you will be introduced to new concepts and find some of them extremely useful. One I found particularly helpful was the notion of *metabolic price.* If a biological brain wants to develop a new cognitive capacity, it must pay a price. The currency in which the price is paid is sugar. Additional energy must be made available and more glucose must be burned to develop and stabilize this new capacity. As in nature in general, there is no such thing as a free lunch. If an animal is to evolve, say, color vision, this new trait must pay by making new sources of food and sugar available to it. If a biological organism wants to develop a conscious self or think in concepts or master a language, then this step into a new level of mental complexity must be sustainable. It requires additional neural hardware, and that hardware requires fuel. That fuel is sugar, and the new trait must enable our animal to find this extra amount of energy in its environment.

Likewise, any good theory of consciousness must reveal how it paid for itself. (In principle, consciousness could be a by-product of other traits that paid for themselves, but the fact that it has remained stable over time suggests that it was adaptive.) A convincing theory must explain how having a world appear to you enabled you to extract more energy from your environment than a zombie could. This evolutionary perspective also helps solve the puzzle of naive realism.

Our ancestors did not need to know that a bear-representation was currently active in their brains or that they were currently attending to an internal state representing a slowly approaching wolf. Thus neither image required them to burn precious sugar. All they needed to know was "Bear over there!" or "Wolf approaching from the left!" Knowing that all of this was just a *model* of the world and of the Now was not necessary for survival. This additional kind of knowledge would have required the formation of what philosophers call *metarepresentations,* or images about other images, thoughts about thoughts. It would have required additional hardware in the brain and more fuel. Evolution sometimes produces superfluous new traits by chance, but these luxurious properties are rarely sustained over long periods of time. Thus, the answer to the question of why our conscious representations of the world are transparent—why we are constitutionally unable to recognize them *as* representations—and why this proved a viable, stable, strategy for survival and procreation probably is that the formation of metarepresentations would not have been cost-efficient: It would have been too expensive in terms of the additional sugar we would have had to find in our environment.

A smaller time scale gives another way of understanding why we were all born as naive realists. Why are we unaware of the tunnel-like nature of consciousness? As noted, the robust illusion of being directly in touch with the outside world has to do with the speed of neural information-processing in our brains. Further, subjective experience is not generated by one process alone but by various interacting functions: multisensory integration, short-term memory, attention, and so on. My theory says that, in essence, consciousness is the *space of attentional agency:* Conscious information is exactly that set of information currently active in our brains to which we can deliberately direct our high-level attention. Low-level attention is automatic and can be triggered by entirely unconscious events. For a perception to be conscious does not mean you deliberately access it with the help of your attentional mechanisms. On the contrary: Most things we're aware of are on the fringe of our consciousness and not in its focus. But whatever is *available* for deliberately directed attention is what is consciously experienced. Nevertheless, if we carefully direct our visual attention at an object, we are constitutionally unable to apprehend the earlier processing stages. "Taking a closer look" doesn't help: We are unable to attend to the construction process that generates the model of the book in our brains. As a matter of fact, attention often seems to do exactly the opposite: by stabilizing the sensory object, we make it even more real.

That is why the walls of the tunnel are impenetrable for us: Even if we believe that something is just an internal construct, we can experience it only as *given* and never as *constructed*. This fact may well be cognitively available to us (because we may have a correct theory or concept of it), but it is not attentionally or introspectively available, simply because on the level of subjective experience, we have no point of reference "outside" the tunnel. Whatever appears to us—however it is mediated—appears as reality.

Please try for a moment to inspect closely the holistic experience of seeing and simultaneously touching the book in your hands and of feeling its weight. Try hard to become aware of the construction process in your brain. You will find two things: First, you cannot do it. Second, the surface of the tunnel is not two-dimensional: It possesses considerable depth and is composed of very different sensory qualities—touch, sound, even smell. In short, the tunnel has a high-dimensional, multi-modal surface. All this contributes to the fact that you cannot recognize the walls of the tunnel as an inner surface; this simply does not resemble any tunnel experience you've ever had.

Why are the walls of the neurophenomenological cave so impenetrable? An answer is that in order to be useful (like the desktop on the graphical user-interface of your personal computer), the inside surface of the cave must be closed and fully realistic. It acts as a dynamic filter. Imagine you could introspectively become aware of ever deeper and earlier phases of your information-processing while looking at the book in your hands. What would happen? The representation would no longer be transparent, but it would still remain inside the tunnel. A flood of interacting patterns would suddenly rush at you; alternative interpretations and intensely competing associations would invade your reality. You would lose yourself in the myriad of micro-events taking place in your brain at every millisecond—you would get lost inside yourself. Your mind would explode into endless loops of self-exploration. Maybe this is what Aldous Huxley meant when, in his 1954 classic, *The Doors of Perception*, he quoted William Blake: "If the doors of perception were cleansed, everything would appear to man as it is, infinite. For man has closed himself up, till he sees all things through narrow chinks of his cavern."

The dynamic filter of phenomenal transparency is one of nature's most intriguing inventions, and it has had far-reaching consequences. Our inner images of the world around us are quite reliable. In order to be good representations, our conscious models of bears, of wolves, of books in our hands, of smiles on our friends' faces, must serve as a window on the world. This window must be clean and crystal clear. That is what phenomenal transparency is: It contributes to the effortlessness and seamlessness that are the hallmark of reliable conscious perceptions that portray the world around us in a sufficiently accurate manner. We don't have to know or care about *how* this series of little miracles keeps unfolding in our brains; we can simply enjoy conscious experience as an invisible interface to reality. As long as nothing goes wrong, naive realism makes for a very relaxed way of living.

However, questions arise. Are there people who aren't naive realists, or special situations in which naive realism disappears? My theory—the self-model theory of subjectivity—predicts that as soon as a conscious representation becomes opaque (that is, as soon as we experience it *as a representation*), we lose naive realism. Consciousness without naive realism does exist. This happens whenever, with the help of other, second-order representations, we become aware of the construction process—of all the ambiguities and dynamical stages preceding the stable state that emerges at the end. When the window is dirty or cracked, we immediately realize that conscious perception is only an interface, and we become aware of the medium itself. We doubt that our sensory organs are working properly. We doubt the existence of whatever it is we are seeing or feeling, and we realize that the medium itself is fallible. In short, if the book in your hands lost its transparency, you would experience it as a state of your mind rather than as an element of the outside world. You would immediately doubt its independent existence. It would be more like a book-thought than a book-perception.

Precisely this happens in various situations—for example, in visual hallucinations during which the patient is aware of hallucinating, or in ordinary optical illusions when we suddenly become aware that we are not in immediate contact with reality. Normally, such experiences make us think something is wrong with our eyes. If you could consciously experience earlier processing stages of the representation of the book in your hands, the image would probably become unstable and ambiguous; it would start to breathe and move slightly. Its surface would become iridescent, shining in different colors at the same time. Immediately you would ask yourself whether this could be a dream, whether there was something wrong with your eyes, whether someone had mixed a potent hallucinogen into your drink. A segment of the wall of the Ego Tunnel would have lost its transparency, and the self-constructed nature of the overall flow of experience would dawn on you. In a nonconceptual and entirely nontheoretical way, you would suddenly gain a deeper understanding of the fact that this world, at this very moment, only *appears* to you.

What if you were born with an awareness of your internal processing? Obviously you would still not be in contact with reality as such, because you would still only know it under a representation. But you would also continuously represent yourself *as representing*. As in a dream in which you have become aware that you're dreaming, your world would no longer be experienced as a reality but as a form of mental content. It would all be one big thought in your mind, the mind of an ideal observer.

We have arrived at a minimalist concept of consciousness. We have an answer to the question of how the brain moves from an internal world-model and an internal Now-model to the full-blown appearance of a world. The answer is this: If the system in which these models are constructed is constitutionally unable to recognize both the world-model and the current psychological moment, the experience of the present, as a model, as only an internal construction, then the system will of necessity generate a reality tunnel. It will have the experience of being in immediate contact with a single, unified world in a single Now. For any such system, *a world appears*. This is equivalent to the minimal notion of consciousness we took as our starting point.

If we can solve the One-World Problem, the Now Problem, and the Reality Problem, we can also find the global neural correlate of consciousness in the human brain. Recall that there is a specific NCC for forms of conscious content (one for the redness of the rose, another for the rose as a whole, and so on) as well as a global NCC, which is a much larger set of neural properties underlying consciousness as a whole, or all currently active forms of conscious content, underpinning your experiential model of the world in its totality at a given moment. Solving the

One-World Problem, the Now Problem, and the Reality Problem involves three steps: First, finding a suitable phenomenological description of what it's like to have all these experiences; second, analyzing their contents in more detail (the representational level); and third, describing the functions bringing about these contents. Discovering the global NCC means discovering how these functions are implemented in the nervous system. This would also allow us to decide which other beings on this planet enjoy the appearance of a world; these beings will have a recognizable physical counterpart in their brains.

On the most simple and fundamental level, the global NCC will be a dynamic brain state exhibiting large-scale coherence. It will be fully integrated with whatever generates the virtual window of presence, because in a sense it *is* this window. Finally, it will have to make earlier processing stages unavailable to high-level attention. I predict that by 2050 we will have found the GNCC, the global neural correlate of consciousness. But I also predict that in the process we will discover a series of technical problems that may not be so easy to solve.

THE INEFFABILITY PROBLEM: WHAT WE WILL NEVER BE ABLE TO TALK ABOUT

Imagine I'm holding color swatches of two similar shades of green up in front of you. There's a difference between the two shades, but it's barely noticeable. (The technical term sometimes used by experts in psychophysics is JND, or "just noticeable difference." The JND is a statistical distinction, not an exact quantity.) The two shades (I'll call them Green No. 24 and Green No. 25) are the nearest possible neighbors on the color chart; there's no shade of green between them that you could discriminate. Now I put my hands behind my back, mix the swatches, and hold one up. Is it Green No. 24 or Green No. 25? The interesting discovery is that conscious perception alone does not enable you to tell the difference. This means that understanding consciousness may also involve understanding the subtle and the ultrafine, not just the whole.

We now must move from the global to the more subtle aspects of consciousness. If it is really true that some aspects of the contents of consciousness are ineffable—and many philosophers, including me, believe this to be the case—how are we going to do solid scientific research on them? How can we reductively explain something we cannot even talk about properly?

The contents of consciousness can be ineffable in many different ways. You cannot explain to a blind man the redness of a rose. If the linguistic community you live in does not have a concept for a particular feeling, you may not be able to discover it in yourself or name it so as to share it with others. A third type of ineffability is formed by all those conscious states ("conscious" because they could in principle be attended to) so fleeting you cannot form a memory trace of them: brief flickers on the fringe of your subjective awareness—perhaps a hardly detectable color change or a mild fluctuation in some emotion, or a barely noticeable glimmer in the mélange of your bodily sensations. There might even be longer episodes of conscious experience—during the dream state, say, or under anesthesia—that are systematically unavailable to memory systems in the brain and that no human being has ever reported. Maybe this is also true of the very last moments before death. Here, however, I'm offering a clearer and better-defined example of ineffability to illustrate the Ineffability Problem.

You can't tell me if the green card I'm holding up is Green No. 24 or Green No. 25. It is well known from perceptual psychology experiments that our ability to discriminate sensory values such as hues greatly exceeds our ability to form direct concepts of them. But in order to talk about this specific shade of green, you need a concept. Using a vague category, like "Some kind of light green," is not enough, because you lose the determinate value, the concrete qualitative *suchness* of the experience.

In between 430 and 650 nanometers, human beings can discriminate more than 150 different wavelengths, or different subjective shades, of color. But if asked to reidentify single colors with a high degree of accuracy, they can do so for fewer than 15.[13] The same is true for other sensory experiences. Normal listeners can discriminate about 1,400 steps of pitch difference across the audible frequency range, but they can recognize these steps as examples of only about 80 different pitches. The University of Toronto philosopher Diana Raffman has stated the point clearly:

"We are much better at discriminating perceptual values (i.e. making same/different judgments) than we are at identifying or recognizing them."[14]

Technically, this means we do not possess *introspective identity criteria* for many of the simplest states of consciousness. Our perceptual memory is extremely limited. You can see and experience the *difference* between Green No. 24 and Green No. 25 if you see both at the same time, but you are unable consciously to represent the *sameness* of Green No. 25 over time. Of course, it may appear to you to be the same shade of Green No. 25, but the subjective experience of certainty going along with this introspective belief is itself appearance only, not knowledge. Thus, in a simple, well-defined way, there is an element of ineffability in sensory consciousness: You can experience a myriad of things in all their glory and subtlety without having the means of reliably identifying them. Without that, you cannot speak about them. Certain experts—vintners, musicians, perfume designers—can train their senses to a much finer degree of discrimination and develop special technical terms to describe their introspective experience. For example, connoisseurs may describe the taste of wine as "connected," "herby," "nutty," or "foxy." Nonetheless, even experts of introspection will never be able to exhaust the vast space of ineffable nuances. Nor can ordinary people identify a match to that beautiful shade of green they saw yesterday. That individual shade is not vague at all; it is what a scientist would call a *maximally determinate value*, a concrete and absolutely unambiguous content of consciousness.

As a philosopher, I like these kinds of findings, because they elegantly demonstrate how subtle is the flow of conscious experience. They show that there are innumerable things in life you can fathom only by experiencing them, that there is a depth in pure perception that cannot be grasped or invaded by thought or language. I also like the insight that *qualia*, in the classic sense coined by Clarence Irving Lewis, never really existed—a point also forcefully made by eminent philosopher of consciousness Daniel C. Dennett.[15] *Qualia* is a term philosophers use for simple sensory experiences, such as the redness of red, the awfulness of pain, the sweetness of peach pie. Typically, the idea was that qualia form recognizable inner essences, irreducible simple properties—the atoms of experience. However, in a wonderful way, this story was too simple— empirical consciousness research now shows us the fluidity of subjective experience, its uniqueness, the irreplaceable nature of the single moment of attention. There are no atoms, no nuggets of consciousness.

The Ineffability Problem is a serious challenge for a scientific theory of consciousness—or at least for finding all its neural correlates. The problem is simply put: To pinpoint the minimally sufficient neural correlate of Green No. 24 in the brain, you must assume your subjects' verbal reports are reliable—that they can correctly identify the phenomenal aspect of Green No. 24 over time, in repeated trials in a controlled experimental setting. They must be able to recognize introspectively the subjectively experienced "suchness" of this particular shade of green— and this seems to be impossible.

The Ineffability Problem arises for the simplest forms of sensory awareness, for the finest nuances of sight and touch, of smell and taste, and for those aspects of conscious hearing that underlie the magic and beauty of a musical experience.[16] But it may also appear for empathy, for emotional and intrinsically embodied forms of communication (see chapter 6 and my conversation with Vittorio Gallese, page 174). Once again, these empirical findings are philosophically relevant, because they redirect our attention to something we've known all along: Many things you can express by way of music (or other art forms, like dance) are ineffable, because they can never become the content of a mental concept or be put into words. On the other hand, if this is so, sharing the ineffable aspects of our conscious lives becomes a dubious affair: We can never be sure if our communication was successful; there is no certainty about *what* actually it was we shared. Furthermore, the Ineffability Problem threatens the comprehensiveness of a neuroscientific theory of consciousness. If the primitives of sensory consciousness are evasive, in the sense that even the experiencing subject possesses no internal criteria to reidentify them by introspection, then we cannot match them with the representational content of neural states—even in principle. Some internal criteria exist, but they are crude: absolutes, such as "pure sweetness," "pure blue," "pure red," and so on. But matching Green No. 24 or Green

No. 25 with their underlying physical substrates in a systematic manner seems impossible, because these shades are just too subtle. If we cannot do the mapping, we cannot do the reduction—that is, arrive at the claim that your conscious experience of Green No. 24 is identical with a certain brain state in your head.

Remember, reduction is a relationship not between the phenomena themselves but between theories. T1 is reduced to T2. One theory—say, about our subjective, conscious experience—is reduced to another—say, about large-scale dynamics in the brain. Theories are built out of sentences and concepts. But if there *are* no concepts for certain objects in the domain of one theory, they cannot be mapped onto or reduced to concepts in the other. This is why it may be impossible to do what most hard scientists in consciousness research would like to do: show that Green No. 24 is identical with a state in your head.

What to do? If identification is not possible, elimination seems to be the only alternative. If the qualities of sensory consciousness cannot be turned into what philosophers call proper *theoretical entities* because we have no identity criteria for them, then the cleanest way of solving the Ineffability Problem may be to follow the path that neurophilosopher Paul Churchland and others suggested long ago—to deny the existence of qualia in the first place. Would the best solution be simply to say that by visually attending to this ineffable shade of Green No. 25 in front of us, we are already directly in touch with a hardware property? That is, what we experience is not some sort of phenomenal representational content but neural dynamics *itself*? In this view, our experience of Green No. 25 would not be a conscious experience at all but instead something physical—a brain state. For centuries, when speaking about "qualities" and color experiences, we were actually misdescribing states of our own bodies, internal states we never recognized as such—the walls of the Ego Tunnel.

We could then posit that if we lack the necessary first-person knowledge, then we must define third-person criteria for these ineffable states. If there are no adequate phenomenological concepts, let's form adequate neurobiological concepts instead. Certainly if we look at the brain dynamics underlying what subjects later describe as their conscious experience of *greenness,* we will observe sameness across time. In principle, we can find objective identity criteria, some mathematical property, something that remains the same in our description connecting the experience of green you had yesterday with the experience you're having right now. And then could we not communicate our inner experiences in neurobiological terms, by saying something like "Imagine the Cartesian product of the experiential green manifold and the Möbius strip of calmness—that is, mildly $K\text{-}314_\gamma$, but moving to $Q\text{-}512_\delta$ and also slightly resembling the 372.509-dimensional shape of Irish moss in norm-space"?

I actually *do* like science fiction. This sci-fi scenario is conceivable, in principle. But are we willing to give up our authority over our own inner states—the authority allowing us to say that these two states *must* be the same because they *feel* the same? Are we willing to hand this epistemological authority over to the empirical sciences of the mind? This is the core of the Ineffability Problem, and certainly many of us would not be ready to take the jump into a new system of description. Because traditional folk-psychology is not only a theory but also a practice, there may be a number of deeper problems with Churchland's strategy of what he calls "eliminative materialism." In his words, "Eliminative materialism is the thesis that our commonsense conception of psychological phenomena constitutes a radically false theory, a theory so fundamentally defective that both the principles and the ontology of that theory will eventually be displaced, rather than smoothly reduced, by completed neuroscience."[17] Churchland has an original and refreshingly different perspective: If we just gave up the idea that we ever had anything like conscious minds in the first place and began to train our native mechanisms of introspection with the help of the new and much more fine-grained conceptual distinctions offered by neuroscience, then we would also *discover* much more, we would *enrich* our inner lives by becoming materialists. "I suggest, then, that those of us who prize the flux and content of our subjective phenomenological experience need not view the advance of materialist neuroscience with fear and foreboding," he has noted. "Quite the contrary. The genuine arrival of a materialist kinematics and dynamics for psychological states and cognitive processes will constitute not a gloom in which our inner life is suppressed or eclipsed, but rather a dawning, in which its marvelous intricacies are finally *revealed*—most notably, if we apply [it] ourselves, in direct self-conscious introspection."[18]

Still, many people would be disinclined to turn something that was previously ineffable into a public property about which they could communicate using the vocabulary of neuroscience. They would feel that this was not what they wanted to know at the outset. More important, they might fear that in pursuit of solving the problem, we had lost something deeper along the way. Theories of consciousness have cultural consequences. I will return to this issue.

THE EVOLUTION PROBLEM: COULDN'T ALL OF THIS HAVE HAPPENED IN THE DARK?

The Evolution Problem is one of the most difficult problems for a theory of consciousness. Why, and in what sense, was it necessary to develop something like consciousness in the nervous systems of animals? Couldn't zombies have evolved instead? Here, the answer is both yes and no.

As I noted in the Introduction, conscious experience is not an all-or-nothing phenomenon; it comes in many shades and flavors. There is a long history of consciousness on this planet. We have strong, converging evidence that all of Earth's warm-blooded vertebrates (and probably certain other creatures) enjoy phenomenal experience. The basic brain features of sensory consciousness are preserved among mammals and exhibit strong homologies due to common ancestry. They may not have language and conceptual thought, but it is likely that they all have sensations and emotions. They are clearly able to suffer. But since they do all this without verbal reports, it is almost impossible to investigate this issue more deeply. What we must understand is how *Homo sapiens* managed to acquire—over the course of our biological history and individually as infants—this amazing property of living our lives in the Ego Tunnel successfully and without realizing it.

First, let's not forget that evolution is driven by chance, does not pursue a goal, and achieved what we now consider the continuous optimization of nervous systems in a blind process of hereditary variation and selection. It is incorrect to assume that evolution *had* to invent consciousness—in principle it could have been a useless by-product. No necessity was involved. Not everything is an adaptation, and even adaptations are not optimally designed, because natural selection can act only on what is already there. Other routes and solutions were and remain possible. Nevertheless, a lot of what happened in our brains and in those of our ancestors clearly was adaptive and had survival value.

Today, we have a long list of potential candidate functions of consciousness: Among them are the emergence of intrinsically motivating states, the enhancement of social coordination, a strategy for improving the internal selection and resource allocation in brains that got too complex to regulate themselves, the modification and interrogation of goal hierarchies and long-term plans, retrieval of episodes from long-term memory, construction of storable representations, flexibility and sophistication of behavioral control, mind reading and behavior prediction in social interaction, conflict resolution and troubleshooting, creating a densely integrated representation of reality as a whole, setting a context, learning in a single step, and so on. It is hard to believe that consciousness should have none of these functions. Consider one example only.

There is a consensus among many leading figures in the consciousness community that at least one of the central functions of phenomenal experience is making information "globally available" to an organism. Bernard Baars's global-workspace metaphor has a functional aspect: Put simply, this theory says that conscious information is that subset of active information in the brain that requires monitoring because it's not clear which of your mental capacities you will need to access this information next. Will you need to direct focal attention at it? Will you need to form a concept of it, to think about it, to report it to other human beings? Will you need to make a flexible behavioral response—one that you have selected and weighed against alternatives? Will you need to link this information to episodic memory, perhaps in order to compare it with things you have seen or heard before? Part of Baars's idea is that you become conscious of something only when you don't know which of the tools in your mental toolbox you'll have to use next.

Note that when you learn a difficult task for the first time, such as tying your shoes or riding a bicycle, your practicing is always conscious. It requires attention, and it takes up many of your resources. Yet as soon as you've mastered tying your shoes or riding a bicycle, you forget all about the learning process—to the point that it becomes difficult to teach the skill to your children. It quickly sinks below the threshold of awareness and becomes a fast and efficient subroutine. But whenever the system is confronted with a novel or challenging stimulus, its global workspace is activated and represented in consciousness. This is also the point when *you* become aware of the process.

Of course, a much more differentiated theory is needed, because there are degrees of availability. Some things in life, such as the ineffable shade of Green No. 25, are available for attention, say, but not for memory or conceptual thought. Other things are available for selective motor control but are accessed so quickly you don't really attend to them: If 100-yard sprinters were to wait until they consciously heard the starter's shot, they would already have lost the race; fortunately, their body hears it before they do. There are many degrees of conscious experience, and the closer science looks, the more blurry the border between conscious and unconscious processing becomes. But the general notion of global availability allows us to tell a convincing story about the evolution of consciousness. Here is my part of the story: Consciousness is a new kind of *organ*.

Biological organisms evolved two different kinds of organs. One kind, such as the liver or the heart, forms part of an organism's "hardware." Organs of this type are permanently realized. Then there are "virtual organs"—feelings (courage, anger, desire) and the phenomenal experience of seeing colored objects or hearing music or having a certain episodic memory. The immune response, which is realized only when needed, is another example of a virtual organ: For a certain time, it creates special causal properties, has a certain function, and does a job for the organism. When the job is done, it disappears. Virtual organs are like physical organs in that they fulfill a specific function; they are coherent assemblies of functional properties that allow you to do new things. Though part of a behavioral repertoire on the macro level of observable traits, they can also be seen as composed of billions of concerted micro-events—immune cells or neurons firing away. Unlike a liver or a heart, they are realized transiently. What we subjectively experience are the processes brought about by the ongoing activity of one or many of such virtual organs.

Our virtual organs make information globally available to us, allowing us to access new facts and sometimes entirely new forms of knowledge. Take as an example the fact that you are holding this book in your hands right now. The phenomenal book (i.e., the conscious book-experience) and the phenomenal hands (i.e., the conscious experience of certain parts of a bodily self) are examples of currently active virtual organs. The neural correlates in your brain work for you as object emulators, internally simulating the book you are holding, without your being aware of the fact. The same is true of the conscious hand-experience, which is part of the bodily subject emulator. The brain is also making other facts available to you: the fact that this book exists, that it has certain invariant surface properties, a certain weight, and so on. As soon as all this information about the existence and properties of the book becomes conscious, it is available for the guidance of attention, for further cognitive processing, for flexible behavior.

Now we can begin to see what the central evolutionary function of consciousness must have been: It makes classes of facts globally available for an organism and thereby allows it to attend to them, to think about them, and to react to them in a flexible manner that automatically takes the overall context into account. Only if a *world appears to you* in the first place can you begin to grasp the fact that an outside reality exists. This is the necessary precondition for discovering the fact that *you* exist as well. Only if you have a consciousness tunnel can you realize that you are part of this reality and are present in it right now.

Moreover, as soon as this global stage—the consciousness tunnel—has been stabilized, many other types of virtual organs can be generated

and begin their dance in your nervous system. Consciousness is an inherently biological phenomenon, and the tunnel is what holds it all together. Within the tunnel, the choreography of your subjective life
begins to unfold. You can experience conscious emotions and thereby
discover that you have certain goals and needs. You can apprehend
yourself as a thinker of thoughts. You can discover that there are other
people—other agents—in the environment and learn about your relationship to them; unless a certain type of conscious experience makes
this fact globally available to you, you cannot cooperate with them, selectively imitate them, or learn from them in other ways. If you are
smart, you may even begin to control their behavior by controlling their
conscious states. If you successfully deceive them—if, say, you manage
to install a false belief in their minds—then you have activated a virtual
organ in another brain.

Phenomenal states are neurocomputational organs that make survival-
relevant information globally available within a window of presence.
They let you become aware of new facts within a unified psychological
moment. Clearly, being able to use all the tools in your mental toolbox
to react to new classes of facts must have been a major adaptive advantage. Every new virtual organ, every new sensory experience, every new
conscious thought had a metabolic price; it was costly to activate them,
if only for a couple of seconds or minutes at a time. But since they paid
for themselves in terms of additional glucose, and in terms of security,
survival, and procreation, they spread across populations and sustain
themselves to this day. They allowed us to discriminate between what
we can eat and what we can't, to search for and detect novel sources of
food, to plan our attack on our prey. They allowed us to read other
people's minds and cooperate more efficiently with our fellow hunters.
Finally, they allowed us to learn from past experience.

The interim conclusion is that making a world appear in an organism's brain was a new computational strategy. Flagging the dangerous
present world as real kept us from getting lost in our memories and our
fantasies. Flagging the present enables a conscious organism to plan
different and more efficient ways of escape or of deceiving or stalking
its prey, namely by comparing internal dry runs of the target behavior

with the features of a given world. If you have a conscious, transparent world-model, you can, for the first time, directly compare what is actual with what is only possible, the actual world with simulated possible worlds you've designed in your mind. High-level intelligence means not only having offline states in which you can simulate potential threats or desired outcomes but also comparing the real situation with a number of possible goal-states. After you have found a path from the real world into the most desirable possible world in your mind, you can begin to act.

It is easy to overlook the causal relevance of this first evolutionary step, the fundamental computational goal of conscious experience. It is the one necessary functional property on which everything else rests. We can simply call it "reality generation": It allowed animals to represent explicitly the fact that something *is actually the case.* A transparent world-model lets you discover that something is really out there, and by integrating your portrait of the world with the subjective Now, it lets you grasp the fact that the world is *present.* This step opened up a new level of complexity. Thus, having a global world-model is a new way of processing information about the world in a highly integrated manner. Every conscious thought, every bodily sensation, every sound and every sight, every experience of empathy or of sharing the goals of another human being makes a different class of facts available for the adaptive, flexible, and selective form of processing that only conscious experience can provide. Whatever is elevated to the level of global availability suddenly becomes more fluid and more context-sensitive and is directly related to all other contents of your conscious mind.

The functions of global availability can be specific: Conscious color vision gives you information about nutritional value, as when you notice the luscious red berries among the green leaves. The conscious experience of empathy provides you with a nonlinguistic form of knowledge about the emotional states of a fellow human being. Once you have this form of awareness, you can attend to it, adapt your motor behavior to it, and associate it with memories of the past. Phenomenal states do not just represent facts about berries or about the feelings of other human beings; they also bind these things into a global processing stage and allow you to use all your mental capacities to explore them further. In short, individual conscious experiences from the object level upward are virtual organs that transiently make knowledge available to you in an entirely new data format—the consciousness tunnel. And your unified global model of a single world provides a holistic frame of reference in which all this can take place.

If a creature such as *Homo sapiens* evolves the additional ability to run offline simulations in its mind, then it can represent possible worlds—worlds that are not experienced as present. This species can have episodic memory. It can develop the ability to plan. It can ask itself, "How would a world look in which I had many children? What would the world be like if I were perfectly healthy? Or if I were rich and famous? And how can I make these things happen? Can I imagine a path leading from the present world into this imagined world?"

Such a being can also enjoy mental time travel, because it can switch back and forth between "inside-time" and "outside-time." It can compare present experiences to past ones—but it can also hallucinate or get lost in its own daydreams. If it wants to use these new mental abilities properly, its brain must come up with a robust and reliable way to tell the difference between representation and simulation. The being must stay anchored in the real world; if you lose yourself in daydreams, sooner or later another animal will come along and eat you. Therefore, you need a mechanism that reliably shows you the difference between the one real world and the many possible ones. And this trick must be achieved on the level of conscious experience itself, which is not an easy problem. As I discussed, conscious experience already *is* a simulation and never brings the subject of experience—you—into direct contact with reality. So the question is, How can you avoid getting lost in the labyrinth of your conscious mind?

A major function of the transparent conscious model of reality is to represent facticity—that is, to generate a rock-bottom frame of reference for the organism using it: something that unfailingly defines what is real (even if it isn't); something you cannot fool around or tamper with. Transparency solved the problem of simulating a multitude of possible inner worlds without getting lost in them; it did so by allowing biological organisms to represent explicitly that one of those worlds is an actual reality. I call this the "world-zero hypothesis."

Human beings know that some of their conscious experiences do not refer to the real world but are only representations in their minds. Now we can see how fundamental this step was, and we can recognize its functional value. Not only were we able to have conscious thoughts, but we could also experience them *as* thoughts, rather than hallucinating or getting lost in a fantasy. This step allowed us to become superbly intelligent. It let us compare our memories and goals and plans with our present situation, and it helped us seek mental bridges from the present to a more desirable reality.

The distinction between things that only appear to us and real, objective facts became an element of our lived reality. (Please note that this is probably not true of most other animals on this planet.) By consciously experiencing some elements of our tunnel as mere images or thoughts about the world, we became aware of the possibility of misrepresentation. We understood that sometimes we can be wrong, since reality is only a specific type of appearance. As evolved representational systems, we could now represent one of the most important facts about ourselves—namely, that we *are* representational systems. We were able to grasp the notions of truth and falsity, of knowledge and illusion. As soon as we had grasped this distinction, cultural evolution exploded, because we became ever more intelligent by systematically increasing knowledge and minimizing illusion.

The discovery of the appearance/reality distinction was possible because we realized that some of the content of our conscious minds is constructed internally and because we could introspectively apprehend the construction process. The technical term here would be *phenomenal opacity*—the opposite of transparency. Those things in the evolution of consciousness that are old, ultrafast, and extremely reliable—such as the qualities of sensory experience—are transparent; abstract conscious thought is not. From an evolutionary perspective, thinking is very new, quite unreliable (as we all know), and so slow that we can actually observe it going on in our brains. In conscious reasoning, we witness the formation of thoughts; some processing stages are available for introspective attention. Therefore, we know that our thoughts are not given but *made*.

The inner appearance of a fully realistic world, as present in the here and now, was an elegant way of creating a frame of reference and a reliable anchor for all those kinds of mental activity necessary for higher forms of intelligence. You can grasp and design possible worlds only if a robust first-order reality is already in place. That was the fundamental breakthrough—as well as the central function of consciousness as such. As it turned out, the consciousness tunnel possessed obvious survival value and was adaptive because it supplied a unified and robust frame of reference for higher levels of reality-modeling. Nevertheless, all this is not even half the story: We need to take one last step up the ladder, a big one. Our brief *tour d'horizon* concludes with the deepest and most difficult puzzle of all: the *subjectivity* of consciousness.

THE WHO PROBLEM: WHAT IS THE ENTITY THAT HAS CONSCIOUS EXPERIENCE?

Consciousness is always bound to an individual first-person perspective; this is part of what makes it so elusive. It is a *subjective* phenomenon. Someone *has* it. In a deep and indisputable way, your inner world truly is not just *someone's* inner world but *your* inner world—a private realm of experience that only you have direct access to.

The conscious mind is not a public object—or such is the orthodox view, which may yet be overthrown by the Consciousness Revolution. In any event, the orthodox view holds that scientific research can be conducted only on objects exhibiting properties that are, at least in principle, observable to all of us. Green No. 24 is not. Neither is the distinct sensory quality of the scent of mixed amber and sandalwood, nor is your empathic experience of understanding the emotions of another human being when you see him in tears. Brain states, on the other hand, are observable. Brain states also clearly have what philosophers call *representational content*. There are receptive fields for the various sensory stimuli. We know where emotional content originates, and we have good candidates for the seat of episodic memory in the brain, and so on.

Conscious experience has content, too—phenomenal content—and I touched upon it in the Introduction: Its phenomenal content is its subjective character—how an experience privately and inwardly *feels* to you, what it is like to have it. But this particular content, it seems, is accessible only to a single person—you, the experiencing subject. And who is that?

To form a successful theory of consciousness, we must match first-person phenomenal content to third-person brain content. We must somehow reconcile the inner perspective of the experiencing self with the outside perspective of science. And there will always be many of us who intuitively think this can never be done. Many people think consciousness is ontologically irreducible (as philosophers say), because first-person facts cannot be reduced to third-person facts. It is more likely, however, that consciousness is epistemically irreducible (as philosophers say). The idea is simple: One reality, one kind of fact, but two kinds of knowledge: first-person knowledge and third-person knowledge. Even though consciousness is a physical process, these two different forms of knowing can never be conflated. Knowing every last thing about a person's brain states will never allow us to know what they are like for the person herself. But the concept of a *first-person perspective* turns out to be vague the moment we take a close look at it. *What is this mysterious first person?* What does the word "I" refer to? If not simply to the speaker, does it refer to anything in the known world at all? Is the existence of an experiencing self a necessary component of consciousness? I don't think it is—for one thing, because there seem to be "self-less" forms of conscious experience. In certain severe psychiatric disorders, such as Cotard's syndrome, patients sometimes stop using the first-person pronoun and, moreover, claim that they do not really exist. M. David Enoch and William Trethowan have described such cases in their book *Uncommon Psychiatric Syndromes:* "Subsequently the subject may proceed to deny her very existence, even dispensing altogether with the use of the personal pronoun 'I'. One patient even called herself 'Madam Zero' in order to emphasize her non-existence. One [patient] said, referring to herself, 'It's no use. Wrap it up and throw "it" in the dustbin."[19]

Mystics of all cultures and all times have reported deep spiritual experiences in which no "self" was present, and some of them, too, stopped using the pronoun "I." Indeed, many of the simple organisms on this planet may have a consciousness tunnel with nobody living in it. Perhaps some of them have only a consciousness "bubble" instead of a tunnel, because, together with the self, awareness of past or future disappears as well.

Note that up to now, in defining the problems for a grand unified theory of consciousness, we have assumed only a minimalist notion: the appearance of a world. But as you are reading these sentences, not only is the light on but there is also somebody home. Human consciousness is characterized by various forms of inwardness, all of which influence one another: First, it is an internal process in the nervous system; second, it creates the experience of being in a world; third, the virtual window of presence gives us temporal internality, a Now. But the deepest form of inwardness was the creation of an internal self/world border.

In evolution, this process started physically, with the development of cell membranes and an immune system to define which cells in one's body were to be treated as one's own and which were intruders.[20] Billions of years later, nervous systems were able to represent this self/world distinction on a higher level—for instance, as body boundaries delineated by an integrated but as yet unconscious body schema. Conscious experience then elevated this fundamental strategy of partitioning reality to a previously unknown level of complexity and intelligence. The phenomenal self was born, and the conscious experience of *being someone* gradually emerged. A self-model, an inner image of the organism as a whole, was built into the world-model, and this is how the consciously experienced first-person perspective developed.

How to comprehend subjectivity is the deepest puzzle in consciousness research. In order to overcome it, we must understand how the conscious self was born into the tunnel, how nature managed to evolve a centered model of reality, creating inner worlds that not only appear but that appear *to someone*. We must understand how the consciousness tunnel turned into an Ego Tunnel.

CHAPTER TWO APPENDIX
THE UNITY OF CONSCIOUSNESS:
A CONVERSATION WITH WOLF SINGER

 **Wolf Singer** is professor of neurophysiology and director of the Department of Neurophysiology at the Max Planck Institute for Brain Research in Frankfurt, Germany. In 2004, he founded the Frankfurt Institute for Advanced Studies (FIAS), which conducts basic theoretical research in various areas of science, bringing together theorists from the disciplines of biology, chemistry, neuroscience, physics, and computer science. His main research interest lies in understanding the neuronal processes underlying higher cognitive functions, such as visual perception, memory, and attention. He is also dedicated to making the results of brain research known to the general public and is a recipient of the Max Planck Prize for Public Science.

Singer has been particularly active in the philosophical debate concerning free will. He is coeditor (with Christoph Engel) of *Better Than Conscious? Decision Making, the Human Mind, and Implications for Institutions* (2008).

Metzinger: Wolf, given the current state of the art, what is the relation between consciousness and feature-binding?

Singer: A unique property of consciousness is its coherence. The contents of consciousness change continuously, at the pace of the experienced present, but at any one moment all the contents of phenomenal awareness are related to one another, unless there is a pathological condition causing a disintegration of conscious experience. This suggests a close relation between consciousness and binding. It seems that only those results of the numerous computational processes that have been bound successfully will enter consciousness simultaneously. This notion also establishes a close link among consciousness, short-term memory, and attention. Evidence indicates that stimuli need to be attended to in order to be perceived consciously, and only then will they have access to short-term memory.

Metzinger: But why is there a binding problem to begin with?

Singer: The binding problem results from two distinct features of the brain: First, the brain is a highly distributed system, in which a very large number of operations are carried out in parallel; second, it lacks a single convergence center, in which the results of these parallel computations could be evaluated in a coherent way. The various processing modules are interconnected, in an exceedingly dense and complex network of reciprocal connections, and these appear to be generating globally ordered states, by means of powerful self-organizing mechanisms. It follows that representations of complex cognitive contents—perceptual objects, thoughts, action plans, reactivated memories—must have a distributed structure as well. This requires that neurons participating in a distributed representation of a particular type of content convey two messages in parallel: First, they have to signal whether the feature they're tuned to is present; second, they have to indicate which of the many other neurons they're cooperating with in forming a distributed representation. It is widely accepted that neurons signal the presence of the feature they encode by increasing their discharge frequency; however, there's less consensus about how neurons signal with which other neurons they cooperate.

Metzinger: What are the constraints for such a signaling?

Singer: Because representations of cognitive contents can change rapidly, it needs to be decipherable with very high temporal resolution. We've proposed that the relation-defining signature is the precise synchronization of the discharges of the individual neurons.

Metzinger: But why synchronization?

Singer: Precise synchronization increases the impact of neuronal discharges, favoring further joint processing of the synchronized messages. Further evidence indicates that such synchronization is best achieved if neurons engage in rhythmic, oscillatory discharges, because oscillatory processes can be synchronized more easily than temporally unstructured activation sequences.

Metzinger: Then this isn't just a hypothesis—there's supportive experimental evidence.

Singer: Since the discovery of synchronized oscillatory discharges in the visual cortex more than a decade ago, more and more evidence has supported the hypothesis that synchronization of oscillatory activity may be the mechanism for the binding of distributed brain processes—whereas the relevant oscillation frequencies differ for different structures and in the cerebral cortex typically cover the range of beta- and gamma-oscillations: 20 to 80 Hz. What makes the synchronization phenomena particularly interesting in the present context is that they occur in association with a number of functions relevant for conscious experience.

Metzinger: Which functions are those?

Singer: These oscillations occur during the encoding of perceptual objects, when coherent representations of the various attributes of these objects have to be formed. The oscillations are consistently observed when subjects direct their attention toward an object and retain information about it in working memory. And finally, the oscillations are a distinctive correlate of conscious perception.

Metzinger: What is the evidence here?

Singer: In a test in which subjects are exposed to stimuli that are degraded by noise so that the stimuli are consciously perceived only half the time, you can study the brain activity selectively associated with conscious experience. Since the physical attributes of the stimuli are the same throughout, you can simply compare brain signals in cases where the subjects consciously perceive the stimuli with the signals in cases where they don't. Investigations reveal that during conscious perception, widely distributed regions of the cerebral cortex transiently engage in precisely synchronized high-frequency oscillations. When the stimuli are not consciously perceived, the various processing regions still engage in high-frequency oscillations—indicating that *some* stimulus-processing is performed—but these are local processes and do not join into globally synchronized patterns. This suggests that access to consciousness requires that a sufficiently large number of processing areas—or in other words, a sufficient number of distributed computations—be bound by synchronization and that those coherent states be maintained over a sufficiently long period.

Metzinger: This could be interesting from a philosophical perspective. Wouldn't this ideally account for the unity of consciousness?

Singer: Indeed, this would also account for the unity of consciousness—for the fact that the contents of phenomenal awareness, although they change from moment to moment, are always experienced as coherent. Admittedly, the argument is somewhat circular, but if it is a necessary prerequisite for access to consciousness that activity be sufficiently synchronized across a sufficient number of processing regions, and if synchronization is equivalent with semantic binding, with integrating the meaning, it follows that the contents of consciousness can only be coherent.

Metzinger: What remains to be shown, if what you describe here turns out to be the case?

Singer: Even if the proposed scenario turns out to be true, the question remains as to whether we have arrived at a satisfactory description of the neuronal correlates of consciousness. What do we gain by saying that the neuronal correlate of consciousness is a particular metastable state of a very complex, highly dynamic, nonstationary distributed system—a state characterized by sequences of ever-changing patterns of precisely synchronized oscillations? Further research will lead to more detailed descriptions of such states—but these will likely be abstract, mathematical descriptions of state vectors. Eventually, advanced analytic methods may reveal the semantic content, the actual meaning of such state vectors, and it may become possible to manipulate these states and thereby alter the contents of consciousness, thus providing causal evidence for the relation between neuronal activity and the contents of phenomenal awareness. However, this is probably about as close as we can come, in our attempts to identify the neuronal correlates of consciousness. How these neuronal activation patterns eventually give rise to subjective feelings, emotions, and so on, will probably remain a conundrum for quite some time even if we arrive at precise descriptions of neuronal states corresponding to consciousness.

Metzinger: In your field, what are the most urgent questions, and where is the field moving?

Singer: The most challenging questions are how information is encoded in distributed neuronal networks and how subjective feelings, the so-called qualia, emerge from distributed neuronal activity. It is commonly held that neurons convey information by modulating their discharge rate—that is, by signaling the presence of contents for which they are specialized through increases in their firing rate. However, accumulating evidence suggests that complex cognitive contents are encoded by the activity of distributed assemblies of neurons and that the information is contained in the relations between the amplitudes and in the duration of the discharges. The great challenge for future work is to extract the information encoded in these high-dimensional time series. This requires simultaneous recordings from a large number of neurons and identification of the relevant spatio-temporal patterns. It is still unclear which aspects of the large number of possible patterns the nervous system exploits to encode information, so searching for these patterns will require developing new and highly sophisticated mathematical search algorithms. Thus, we'll need close collaboration between experimentalists and theoreticians to advance our understanding of the neuronal processes underlying higher cognitive functions.

Metzinger: Wolf, why are you so interested in philosophy, and what kind of philosophy would you like to see in the future? What relevant contributions from the humanities are you waiting for?

Singer: My interest in philosophy is nurtured by the evidence that progress in neurobiology will provide some answers to the classic questions treated in philosophy. This is the case for epistemology, philosophy of mind, and moral philosophy. Progress in cognitive neuroscience will tell us how we perceive and to what extent our perceptions are reconstructions rather than representations of absolute realities. As we learn more about the emergence of mental functions from complex neuronal interactions, we will gain insight into possible solutions of the mind-body problem, and as we learn to understand how our brains assign values and distinguish between appropriate and inappropriate conditions, we will learn more about the evolution and constitution of morality.

Conversely, cognitive neuroscience needs the humanities—for several reasons. First, progress in the neurosciences raises a large number of new ethical problems, and these need to be addressed not only by neurobiologists but also by representatives of the humanities. Second, as neuroscience progresses, more and more phenomena that have traditionally been the subject of humanities research can be investigated with neuroscientific methods; thus, the humanities will provide the taxonomy and description of phenomena awaiting investigation at the neuronal level. Brain research begins with the analysis of such phenomena as empathy, jealousy, altruism, shared attention, and social imprinting—phenomena that have traditionally been described and analyzed by psychologists, sociologists, economists, and philosophers. Classification and precise description of these phenomena are prerequisites for the neuroscientific attempts to identify the underlying neuronal processes. There will undoubtedly be close collaborations in the near future between the neurosciences and the humanities—a fortunate development, as it promises to overcome some of the dividing lines that have segregated the natural sciences from the humanities over the last centuries.

PART TWO

IDEAS AND DISCOVERIES

THREE

✦

OUT OF THE BODY AND INTO THE MIND

Body Image, Out-of-Body Experiences, and the Virtual Self

"Owning" your body, its sensations, and its various parts is fundamental to the feeling of *being someone.* Your body image is surprisingly flexible. Expert skiers, for example, can extend their consciously experienced body image to the tips of their skis. Race-car drivers can expand it to include the boundaries of the car; they do not have to judge visually whether they can squeeze through a narrow opening or avoid an obstacle—they simply feel it. Have you ever tried to walk with your eyes closed, or in the dark, tapping ahead with a stick as a blind person does? If so, you've probably noticed that you suddenly start to feel a tactile sensation at the end of the stick. All these are examples of what philosophers call the *sense of ownership,* which is a specific aspect of conscious experience—a form of automatic self-attribution that integrates a certain kind of conscious content into what is experienced as one's self.

Neuroimaging studies have given us a good first idea of what happens in the brain when the sense of ownership, as illustrated by the rubber-hand experiment discussed in the Introduction, is transferred from a subject's real arm to the rubber hand: Figure 2 shows areas of increased activity in the premotor cortex. It is plausible to assume that at the moment you consciously experience the rubber hand as part of

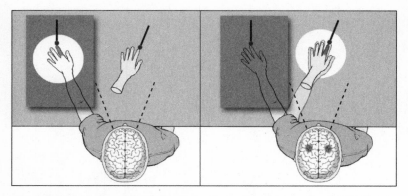

Figure 2: The rubber-hand illusion. The illustration on the right shows the subject's illusion as the felt strokes are aligned with the seen strokes of the probe. The dark areas show heightened activity in the brain; the phenomenally experienced, illusory position of the arm is indicated by the light outline. The underlying activation of neurons in the premotor cortex is demonstrated by experimental data. (Botvinick & Cohen, "Rubber Hand 'Feels' Touch," ibid.)

your body, a fusion of the tactile and visual receptive fields takes place and is reflected by the activation of neurons in the premotor cortex.[1]

The rubber-hand illusion helps us understand the interplay among vision, touch, and proprioception, the sense of body posture and balance originating in your vestibular system. Your bodily self-model is created by a process of multisensory integration, based on a simple statistical correlation your brain has discovered. The phenomenal incorporation of the rubber hand into your self-model results from correlated tactile and visual inputs. As the brain detects the synchronicity underlying this correlation, it automatically forms a new, coherent representation. The consciously experienced sense of ownership follows.

In Matthew Botvinick and Jonathan Cohen's study, subjects were asked to close their eyes and point to their concealed left hand; they tended to point in the direction of the rubber one, with the degree of mispointing dependent on the reported duration of the illusion. In a similar experiment, conducted by K. C. Armel and V. S. Ramachandran at UCSD's Brain and Perception Laboratory, if one of the rubber fingers was bent backward into a physiologically impossible position, subjects not only experienced their phenomenal finger as being bent but also exhibited a significant skin-conductance reaction, indicating that unconscious autonomous mechanisms, which cannot be controlled at will, were also reacting to the assumption that the rubber hand was part of the self. Only two out of one hundred and twenty subjects reported feeling actual pain, but many pulled back their real hands and widened their eyes in alarm or laughed nervously.[2]

The beauty of the rubber-hand illusion is that you can try it at home. It clearly shows that the consciously experienced sense of ownership is directly determined by representational processes in the brain. Note how, in your subjective experience, the transition from shoulder to rubber hand is seamless. Subjectively, they are both part of one and the same bodily self; the quality of "ownership" is continuous and distributed evenly between them. You don't need to do anything to achieve this effect. It seems to be the result of complex, dynamic self-organization in the brain. The emergence of the bodily self-model—the conscious image of the body as a whole—is based on a subpersonal, automatic process of binding different features together—of achieving coherence. This coherent structure is what you experience as your own body and your own limbs.

There are a number of intriguing further facts—such as the finding that subjects will mislocate their real hand only when the rubber one is in a physiologically realistic position. This indicates that "top-down" processes, such as expectations about body shape, play an important role. For example, a principle of "body constancy" seems to be at work, keeping the number of arms at two. The rubber hand displaces the real hand rather than merely being mistaken for it. Recently, psychometric studies have shown that the feeling of having a body is made up of various subcomponents—the three most important being *ownership,* *agency,* and *location*—which can be dissociated.[3] "Me-ness" cannot be reduced to "here-ness," and, more important, agency (that is, the performance of an action) and ownership are distinct, identifiable, and separable aspects of subjective experience. Gut feelings ("interoceptive body perception") and background emotions are another important cluster anchoring the conscious self,[4] but it is becoming obvious that ownership is closest to the core of our target property of selfhood. Nevertheless, the experience of being an embodied self is a holistic construct, characterized by part-whole relationships and stemming from many different sources.[5]

Phenomenal ownership is not only at the heart of conscious self-experience; it also has unconscious precursors. Classical neurology hypothesized about a *body schema*, an unconscious but constantly updated brain map of limb positions, body shape, and posture.[6] Recent research shows that Japanese macaques can be trained to use tools even though they only rarely exhibit tool use in their natural environment.[7] During successful tool use, changes occur in specific neural networks in their brains, a finding suggesting that the tools are temporarily integrated into their body schemata. When a food pellet is dispensed beyond their reach and they use a rake to bring it closer, a change is observed in their bodily self-model in the brain. In fact, it looks as though their model of their hand and of the space around it is extended to the tip of the tool; that is, on the level of the monkey's model of reality, properties of the hand are transferred to the tool's tip. Certain visual receptive fields now extend from a region just beyond the fingertips to the tip of the rake the monkey is holding, because the parietal lobe in its brain has temporarily incorporated the rake into the body model. In human beings, repeated practice can turn the tip of a tool into a part of the hand, and the tool can be used as sensitively and as skillfully as the fingers.

Recent neuroscientific data indicate that any successful extension of behavioral space is mirrored in the neural substrate of the body image in the brain. The brain constructs an internalized image of the tool by assimilating it into the existing body image. Of course, we do not know whether monkeys actually have the conscious experience of ownership or only the unconscious mechanism. But we do know of several similarities between macaques and human beings that make plausible the assumption that the macaques' morphed and augmented bodily self is conscious.

One exciting aspect of these new data is that they shed light on the evolution of tool use. A necessary precondition of expanding your

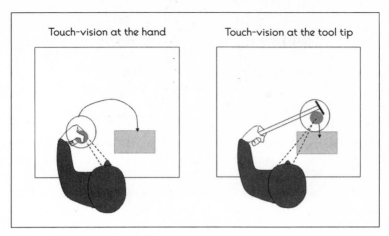

Figure 3: Integrating touch and sight. The subject tries to move a coin (small dark circle) onto a tray with her own hand and with the help of a tool. In the figure on the right, the integrated experience of vision and touch is transferred from the hand to the tip of a tool. The dotted lines trace the subject's gaze direction; the arrows indicate the direction of the movement. The large white circle shows the area where—according to the conscious model of reality—the sense of sight and the sense of touch are integrated. Figure courtesy of Angelo Maravita.

space of action and your capabilities by using tools clearly seems to be the ability to integrate them into a preexisting self-model. You can engage in goal-directed and intelligent tool use only if your brain temporarily represents the tools as part of your self. Intelligent tool use was a major achievement in human evolution. One can plausibly assume that some of the elementary building blocks of human tool-use abilities existed in the brains of our ancestors, 25 million years ago. Then, due to some not-yet-understood evolutionary pressure, they exploded into what we see in humans today.[8] The flexibility in the monkey's body schema strongly relies on properties of body maps in its parietal lobe. The decisive step in human evolution might well have been making a larger part of the body model globally available—that is, accessible to conscious experience. As soon as you can consciously experience a tool as integrated into your bodily self, you can also attend to this process, optimize it, form concepts about it, and control it in a more fine-grained manner—performing what today we call *acts of will*. Conscious self-experience clearly is a graded phenomenon; it increases in strength as an organism becomes more and more sensitive to an internal context and expands its capacities for self-control.

Monkeys also seem able to incorporate into their bodily self-model a visual image of their hand as displayed on a computer monitor. If an image of a snake or a spider approaches the image of the hand on the screen, the animal retracts its real hand. Monkeys can even learn to control a brain-machine interface that lets them grasp objects with a robot arm controlled by certain parts of their brain.[9] Perhaps most exciting from a philosophical perspective is the idea that all of this may have contributed to the evolutionary emergence of a quasi-Cartesian "meta-self," the capacity to distance yourself from your bodily self—namely, by beginning to see your own body as a tool.[10]

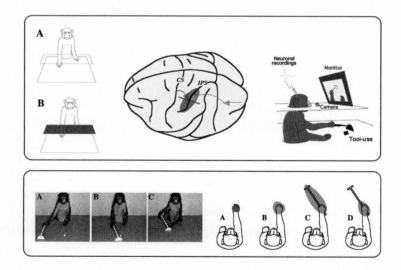

Figure 4: Japanese macaques exhibit intelligent tool use. They can use a rake to reach a food pellet (bottom), and they can monitor their own movements with the help of images on a computer screen, even when their hands are invisible (middle and top): A mere extension of behavioral space, or an extension of the phenomenal self-model? Figures courtesy of Atsushi Iriki.

Clearly, the visual image of the robot arm, just as in the rubber-hand illusion, is embedded in the dancing self-pattern in the macaque's brain. The integration of feedback from the robot arm into this self-model is what allows the macaque to control the arm—to incorporate it function-ally into a behavioral repertoire. In order to develop intelligent tool use, the macaque first had to embed this rake in its self-model; otherwise, it could not have understood that it could use the rake as an extension of its body. There is a link between selfhood and extending global control.

Human beings, too, treat virtual equivalents of their body parts as seen on a video screen as extensions of their own bodies. Just think of mouse pointers on computer desktops or controllable fantasy figures in video games. This may explain the sense of "presence" we sometimes have when playing these ultrarealistic games. Incorporation of artificial actuators into widely distributed brain regions may someday allow hu-man patients successfully to operate advanced prostheses (which, for example, send information from touch and position sensors to a brain-implanted, multichannel recording device via a wireless link), while also enjoying a robust conscious sense of ownership of such devices. All of this gives us a deeper understanding of ownership. On higher levels, ownership is not simply passive integration into a conscious self-model: More often it has to do with functionally integrating something into a feedback loop and then making it part of a control hierarchy. It now looks as if even the evolution of language, culture, and abstract thought might have been a process of "exaptation," of using our body maps for new challenges and purposes—a point to which I return in the chapter on empathy and mirror neurons. Put simply, exaptation is a shift of function for a certain trait in the process of evolution: Bird feathers are a classic example, because initially these evolved "for" temperature regu-lation but later were adapted for flight. Here, the idea is that having an integrated bodily self-model was an extremely useful new trait because it made a host of unexpected exaptations possible.

Clearly, a single *general* mechanism underlies the rubber-hand illu-sion, the evolution of effortless tool use, the ability to experience bodily presence in a virtual environment, and the ability to control artificial de-vices with one's brain. This mechanism is the self-model, an integrated representation of the organism as a whole in the brain. This representation is an ongoing process: It is flexible, can be constantly updated, and allows you to own parts of the world by integrating them into it. Its content is the content of the Ego.

THE OUT-OF-BODY EXPERIENCE

My own interest in consciousness arose from a variety of sources, which were mostly academic but also autobiographical. At some points, the theoretical problem appeared directly and unexpectedly in my life. As a young man, I encountered a series of disturbing experiences, of which the following is a typical instance:

> It is spring, 1977. I am nineteen years old. I am lying in bed, on my back, going to sleep, deeply relaxed yet still alert. The door is half open, and light seeps in. I hear my family's voices from the hallway and the bathroom and pop music from my sister's bedroom. Suddenly I feel as though my bed is sliding into a vertical position, with the head of the bed moving toward the ceiling. I seem to leave my physical body, rising slowly into an upright position. I can still hear the voices, the sound of people brushing their teeth, and the music, but my sight is somewhat blurred. I feel a mixture of amazement and rising panic, sensations that eventually lead to something like a faint, and I find myself back in bed, once again locked into my physical body.

This brief episode was startling for its clarity, its crisp and lucid quality, and the fact that from my point of view it appeared absolutely real. Six years later, I was aware of the concept of the out-of-body experience (OBE), and when such episodes occurred, I could control at least parts of the experience and attempt to make some verifiable observations. As I briefly pointed out in the Introduction, OBEs are a well-known class of states in which one undergoes the highly realistic illusion of leaving one's physical body, usually in the form of an etheric double,

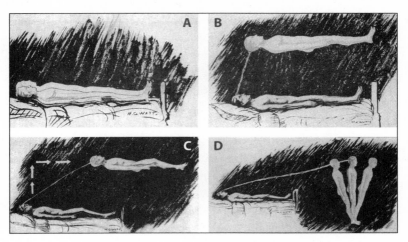

Figure 5: Kinematics of the phenomenal body image during OBE onset: The "classical" motion pattern according to S. Muldoon and H. Carrington, *The Projection of the Astral Body* (London: Rider & Co., 1929).

and moving outside of it. Most OBEs occur spontaneously, during sleep onset or surgical operations or following severe accidents. The classic defining characteristics include a visual representation of one's body from a perceptually impossible, third-person perspective (for example, lying on the bed below) plus a second representation of one's body, typically hovering above.

At about the same time, in the early 1980s, I underwent an equally disturbing experience in my intellectual life. I was writing my philosophy dissertation at Johann-Wolfgang-Goethe University on the discussion of the so-called mind-body problem that ensued after Gilbert Ryle's 1949 book, *The Concept of Mind*. In this period, various philosophers, from Ullin T. Place to Jaegwon Kim, had developed nearly a dozen major theoretical proposals to solve that age-old puzzle, and great progress had been made. I had grown up in a more traditionally oriented philosophy department, which was dominated by the political philosophy of the Frankfurt School. There, almost no one seemed aware of the enormous progress in analytical philosophy of mind. To my great surprise, I discovered that in the really convincing, substantial work at the research frontier, materialism had long ago become the orthodoxy. Almost no one seemed even remotely to consider the possibility of the existence of a soul. There were very few dualists—except on the Continent. It was sobering to realize that some forty years after the end of World War II, with practically all of the German-Jewish intelligentsia and other intellectuals either murdered or driven into exile, many lines of tradition and teacher-student relationships were severed, and German philosophy had been largely decoupled from the global context of discussion. Most German philosophers would not read what was being published in English. Suddenly some of the philosophical debates I witnessed in German universities increasingly struck me as badly informed, a bit provincial, and lacking awareness of where humankind's great project of constructing a comprehensive theory of mind actually stood. I gradually became convinced, by my own reading, that indeed there was no convincing empirical evidence of conscious experience possibly taking place outside the brain, and that the general trend at the frontier of the very best of philosophy of mind clearly pointed in the opposite direction. On the other hand, I had myself repeatedly experienced leaving my body—vividly and in a crystal-clear way. What to do?

There was only one answer: I had to turn these episodes into a controllable and repeatable state of consciousness, and I had to settle experimentally the issue of whether it was possible to make verifiable observations in the out-of-body state. I read everything on OBEs I could find, and I tried various psychological techniques to produce the phenomenon deliberately. In a series of pitiless self-experiments, I stopped drinking liquids at noon, stared at a glass of water by the kitchen sink with the firm intention of returning to it in the out-of-body state, and went to bed thirsty with half a tablespoon of salt in my cheek (you can try this at home). In the scientific literature, I had read that OBEs were associated with the anesthetic ketamine. So when I had to undergo minor surgery in 1985, I talked the anesthetist into changing the medication so that I could experience the wake-up phase of ketamine-induced anesthesia in a medically controlled, experimental setting. (*Do not* try this at home!) Both types of research projects failed, and I gave up on them many years ago. I was never able to go beyond pure first-person phenomenology—that is, to make a single verifiable observation in the OBE state that could even remotely count as evidence for the genuine separability of consciousness and the brain.

In some of my recent research, I have been trying to disentangle the various layers of the conscious self-model—of the Ego. I firmly believe that, from a theoretical perspective, it is most important first to isolate clearly the *simplest* form of self-consciousness. What is the most fundamental, the earliest sense of selfhood? Can we subtract thinking, feeling, and autobiographical memory and still have an Ego? Can we remain in the Now, perhaps even without any acts of will and in the absence of any bodily behavior, and still enjoy phenomenal selfhood? Philosophers in the past have made the mistake of almost exclusively discussing high-level phenomena such as mastery of the first-person pronoun "I" or cognitively mediated forms of intersubjectivity. I contend that we must pay attention to the causally enabling and necessary low-level details first, to what I call "minimal phenomenal selfhood";[11] we must ground the self, and we must do it in an interdisciplinary manner. As you will see, OBEs are a perfect entry point.

Not too long ago, OBEs were something of a taboo zone for serious researchers, just as consciousness was in the early 1980s; both have been considered career-limiting moves by junior researchers. But after decades of neglect, OBEs have now become one of the hottest topics in research on embodiment and the conscious self. Olaf Blanke, whom we met in the Introduction, and I are studying the experience of disembodiment in order to understand what an embodied self truly is.

From a philosophical perspective, OBEs are interesting for a number of reasons. The phenomenology of OBEs inevitably leads to dualism and to the idea of an invisible, weightless, but spatially extended second body. I believe this may actually be the folk-phenomenological ancestor of the notion of a "soul" and of the philosophical protoconcept of the mind.[12] The soul is the OBE-PSM. The traditional concept of an immortal soul that exists independently of the physical body probably has a recent neurophenomenological correlate. In its origins, the "soul" may have been not a metaphysical notion but simply a phenomenological one: the content of the phenomenal Ego activated by the human brain during out-of-body experiences.

In the history of ideas, contemporary philosophical and scientific debates about the mind developed from this protoconcept—an animist, quasi-sensory theory about what it means to have a mind. Having a mind meant having a soul, an ethereal second body. This mythical idea of a "subtle body" that is independent of the physical body and is the carrier of higher mental functions, such as attention and cognition, is found in many different cultures and at many times—for instance, in prescientific theories about a "breath of life."[13] Examples are the Hebrew *ruach*, the Arabic *ruh*, the Latin *spiritus*, the Greek *pneuma*, and the Indian *prana*. The subtle body is a spatially extended entity that was said to keep the physical body alive and leave it after death.[14] It is also known in theosophy and in other spiritual traditions; for instance, as "the resurrection body" and "the glorified body" in Christianity, "the most sacred body" and "supracelestial body" in Sufism, "the diamond body" in Taoism and Vajrayana, "the light body" or "rainbow body" in Tibetan Buddhism.

My theory—the self-model theory of subjectivity—says that this subtle body does indeed exist, but it is not made of "angel stuff" or "astral matter." It is made of pure information, flowing in the brain.[15] Of course, the "flow of information" is just another metaphor, but the information-processing level of description is the best we have at this stage of research. It creates empirically testable hypotheses, and it allows us to see things we could not see before. *The subtle body is the brain's self-model,* and scientific research on the OBE shows this in a particularly striking way.

First-person reports of OBEs are available in abundance, and they, too, come from all times and many different cultures. I propose that the functional core of this kind of conscious experience is formed by a culturally invariant neuropsychological potential common to all human beings. Under certain conditions, the brains of all human beings can generate OBEs. We are now beginning to understand the properties of the functional and representational architecture involved. Examining the phenomenology in OBE reports will help us to understand not only these properties as such but also their neural implementation. There may well be a spatially distributed but functionally distinct neural correlate for the OBE state. In her work, the psychologist Susan J. Blackmore has propounded a reductionist theory of out-of-body experiences, describing them as models of reality created by brains that are cut off from sensory input during stressful situations and have to fall back on internal sources of information.[16] She drew attention to the remarkable fact that visual cognitive maps reconstructed from memory are most often organized from a bird's-eye perspective. Close your eyes and remember the last time you were walking along the beach. Is your visual memory one of looking out at the scene itself? Or is it of observing yourself, perhaps from somewhere above, walking along the coastline? For most people, the latter is the case.

When I first met Blackmore, in Tübingen in 1985, and told her about several out-of-body experiences of my own, she kept asking me to describe, painstakingly, how I moved during these episodes. Not until then did I realize that when I moved around my bedroom at night in the OBE state, it was not in a smooth, continuous path, as in real-life walking or as one might fly in a dream. Instead, I moved in "jumps"—say, from one window to the next. Blackmore has hypothesized that during OBEs we move in discrete shifts, from one salient point in our cognitive map to the next. The shifts take place in an internal model of our environment—a coarse-grained internal simulation of landmarks in settings with which we are familiar. Her general idea is that the OBE is a conscious simulation of the world—spatially organized from a third-person perspective and including a realistic representation of one's own body—and it is highly realistic because we do not recognize it as a simulation.[17]

Blackmore's theory is interesting because it treats OBEs as behavioral spaces. And why shouldn't they be internally simulated behavioral spaces? After all, conscious experience itself seems to be just that: an inner representation of a space in which perceptions are meaningfully integrated with one's behavior. What I found most convincing about Blackmore's OBE model were the jumps from landmark to landmark, a phenomenological feature I had overlooked in my own OBE episodes.

My fifth OBE was particularly memorable. It took place at about 1:00 A.M., on October 31, 1983:

My vision was generally poor during OBE experiences, as would be expected in a dark bedroom at night. When I realized I was unable to flip the light switch in front of which I found myself standing in my OBE state, I became extremely nervous. In order not to ruin everything and lose a precious opportunity for experiments, I decided to stay put until I had calmed down. Then I attempted to walk to the open window, but instead found myself smoothly gliding there, arriving almost instantaneously. I carefully touched the wooden frame, running my hands over it. Tactile sensations were clear but different—that is, the sensation of relative warmth or cold was absent. I leaped through the window and went upward in a spiral. A further phenomenological feature accompanied this experience—the compulsive urge to visualize the headline in the local newspapers: "Was it attempted suicide or an extreme case of somnambulism? Philosophy student drops to his death after sleepwalking out the window." A bit later, I was lying on top of my physical body in bed again, from which I rose in a controlled fashion, for the second time now. I tried to fly to a friend's house in Frankfurt, eighty-five kilometers away, where I intended to try to make some verifiable observations. Just by concentrating on my destination, I was torn forward at great speed, through the wall of my bedroom, and immediately lost consciousness. As I came to, half-locked into my physical body, I felt my clarity decreasing and decided to exit my body one last time.

These incidents, taken from what was a more comprehensive experience, demonstrate a frequently overlooked characteristic of self-motion in the OBE state—namely, that the body model does not move as the physical body would, but that often merely thinking about a target location gets you there on a continuous trajectory. Vestibulo-motor sensations are strong in the OBE state (indeed, one fruitful way of looking at OBEs is as complex vestibulo-motor hallucinations), but weight sensations are only weakly felt, and flying seems to come naturally as the logical means of OBE locomotion. Because most OBEs happen at night, another implicit assumption is that you cannot see very well. That is, if you are jumping from one landmark in your mental model of reality to the next, it is not surprising that the space between two such salient points is experientially vague or underdetermined; at least I simply didn't *expect* to see much detail. Note that the absence of thermal sensations and the short blackouts between different scenes are also well documented in dream research (see chapter 5).

Here are some other first-person accounts of OBEs. This one comes from Swiss biochemist Ernst Waelti, who conducts research at the University of Bern's Institute of Pathology on virosomes for drug delivery and gene transfer:

> I awoke at night—it must have been about 3 A.M.—and realized I was unable to move. I was absolutely certain I was not dreaming, as I was enjoying full consciousness. Filled with fear about my current condition, I had only one goal—namely, to be able to move my body again. I concentrated all my will power and tried to roll over onto my side: Something rolled, but not my body— something that was me, my whole consciousness, including all of its sensations. I rolled onto the floor beside the bed. While this was happening, I did not feel bodiless but as if my body consisted of a substance constituted of a mixture of gas and liquid. To this day, I have not forgotten the amazement that gripped me when I felt myself falling to the floor, but the expected hard impact never came. Had my normal body fallen like that, my head would have collided with the edge of my bedside table. Lying on the floor, I was seized by panic. I knew I possessed a body, and I had only one overwhelming desire: to be able to control it again. With a sudden jolt, I regained control of it, without knowing how I managed to get back into it.

Again from Waelti, about another occasion:

> In a dazed state, I went to bed at 11 P.M. and tried to fall asleep. I was restless and turned over frequently, causing my wife to grumble briefly. Now I forced myself to lie in bed motionless. For a while, I dozed, then felt the need to move my hands, which

were lying on the blanket, into a more comfortable position. In the same instant, I realized that . . . my body was lying there in some kind of paralysis. Simultaneously, I found I could pull my hands out of my physical hands, as if the latter were just a stiff pair of gloves. The process of detachment started at the finger-tips, in a way that could be felt clearly, with a perceptible sound, a kind of crackling. This was precisely the movement I had in-tended to carry out with my physical hands. With this, I de-tached from my body and floated out of it head first, attaining an upright position, as if I were almost weightless. Nevertheless, I had a body, consisting of real limbs. You have certainly seen how elegantly a jellyfish moves through the water. I could now move around with the same ease.

I lay down horizontally in the air and floated across the bed, like a swimmer who has pushed himself off the edge of a swim-ming pool. A delightful feeling of liberation arose within me. But soon I was seized by the ancient fear common to all living creatures—the fear of losing my physical body. It sufficed to drive me back into my body.[18]

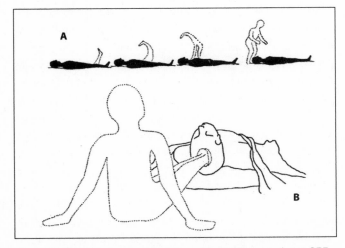

Figure 6 a & b: How the conscious image of the body moves during OBE onset. Two alternative but equally characteristic motion patterns as described by Swiss biochemist Ernst Waelti (1983).

As noted, the sleep paralysis Waelti describes is not a necessary condition for OBEs. They frequently occur following accidents, in combat situations, or during extreme sports—for instance in high-altitude climbers or marathon runners:

> A Scottish woman wrote that, when she was 32 years old, she had an OBE while training for a marathon. "After running approximately 12–13 miles . . . I started to feel as if I wasn't looking through my eyes but from somewhere else. . . . I felt as if something was leaving my body, and although I was still running along looking at the scenery, I was looking at myself running as well. My 'soul' or whatever, was floating somewhere above my body high enough up to see the tops of the trees and the small hills."[19]

Various studies show that between 8 and 15 percent of people in the general population have had at least one OBE.[20] There are much higher incidences in certain groups of people, such as students (25 percent), paranormal believers (49 percent), and schizophrenics (42 percent); there are also OBEs of neurological origin, as in epileptics.[21]

> A 29-year-old woman has had absence seizures since the age of 12 years. The seizures occur five times a week without warning. They consist of a blank stare and brief interruption of ongoing behavior, sometimes with blinking. She had an autoscopic experience at age 19 years during the only generalized tonoclonic seizure she has ever had. While working in a department store she suddenly fell, and she said, "The next thing I knew I was floating just below the ceiling. I could see myself lying there. I wasn't scared; it was too interesting. I saw myself jerking and overheard my boss telling someone to 'punch the timecard out' and that she was going with me to the hospital. Next thing, I was in space and could see Earth. I felt a hand on my left shoulder, and when I went to turn around, I couldn't. Then I looked down and I had no legs; I just saw stars. I stayed there for a while until some inner voice told me to go back to the body. I didn't want to

go because it was gorgeous up there, it was warm—not like heat, but security. Next thing, I woke up in the emergency room." No abnormalities were found on the neurological examination. Skull CT scan was normal. The EEG demonstrated generalized bursts of 3/s spike-and-wave discharges. [22]

At first, the realistic quality of these OBEs seems to argue against their hallucinatory nature. More interesting, though, is how veridical elements and hallucination are integrated into a single whole. Often, the appearance/reality distinction is available: There is insight, but this insight is only partial. One epileptic patient noted that his body, perceived from an external perspective, was dressed in the clothes he was really wearing, but, curiously, his hair was combed, though he knew it had been uncombed before the onset of the episode. Some epileptic patients report that their hovering body casts a shadow; others do not report seeing the shadow. For some, the double is slightly smaller than life-size. We can also see the insight component in the first report by Ernst Waelti previously quoted: "Had my normal body fallen like that, my head would have collided with the edge of my bedside table."

Another reason the OBE is interesting from a philosophical perspective is that it is the best known state of consciousness in which two self-models are active at the same time. To be sure, only one of them is the "locus of identity," the place where the *agent* (in philosophy, an entity that acts) resides. The other self-model—that of the physical body lying, say, on the bed below—is not, strictly speaking, a self-model, because it does not function as the origin of the first-person perspective. This second self-model is not a subject model. It is not the place from which you direct your attention. On the other hand, it is still your own body that you are looking at. You recognize it as your own, but now it is not the body *as subject,* as the locus of knowledge, agency, and conscious experience. That is exactly what the Ego is. These observations are interesting because they allow us to distinguish different functional layers in the conscious human self.

Interestingly, there is a range of phenomena of autoscopy (that is, the experience of viewing your body from a distance) that are probably functionally related to OBEs, and they are of great conceptual interest.

The four main types are autoscopic hallucination, heautoscopy, out-of-body experience, and the "feeling of a presence." In autoscopic hallucinations and heautoscopy, patients see their own body outside, but they do not identify with it and don't have the feeling that they are "in" this illusory body. However, in heautoscopy, things may sometimes go back and forth, and the patient doesn't know which body he is in right now. The shift in the visuospatial first-person perspective, localization, and identification of the self with an illusory body at an extracorporeal position are complete in out-of-body experiences. Here the self and the visuospatial first-person perspective are localized outside one's body, and people see their physical body from this disembodied location. The "feeling of a presence"—which has also been caused by directly stimulating the brain with an electrode—is particularly interesting: It is not a visual own-body illusion but an illusion during which a second illusory body is only *felt* (but not seen).[23]

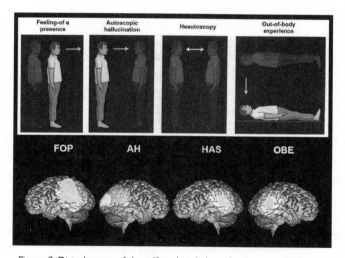

Figure 7: Disturbances of the self and underlying brain areas. All these phenomena show that not only identification with and localization of body parts but also the conscious representation of the entire body and the associated sense of selfhood can be disturbed. All four types of experience are caused by multisensory disintegration having a clear-cut neurological basis (see light areas); brain tumors and epilepsy are among the most frequent causes for heautoscopy. Modified from O. Blanke; Illusions visuelles. In A.B. Safran, A. Vighetto, T. Landis, E. Cabanis (eds.), *Neurophtalmologie* (Paris: Madden, 2004), 147–150.

What about personality correlates? Differential psychology has shown that significant personality traits of people who frequently experience OBEs include openness to new experience, neuroticism, a tendency toward depersonalization (an emotional disorder in which there is loss of contact with one's own personal reality, accompanied by feelings of unreality and strangeness; often people feel that their body is unreal, changing, or dissolving), schizotypy (sufferers experience distorted thinking, behave strangely, typically have few, if any, close friends, and feel nervous around strangers), borderline personality disorder, and histrionics.[24] Another recent study links OBEs to a capacity for strong absorption—that is, experiencing the phenomenal world, in all its aspects and with all one's senses, in a manner that totally engages one's attention and interest—and somatoform dissociation (in part, a tendency to cut one's attention off from bodily and motion stimuli), and points out that such experiences should not automatically be construed as pathological.[25]

It is also interesting to take a closer look at the *phenomenology* of OBEs. For example, the "head exit" depicted in figure 6a is found in only 12.5 percent of cases. The act of leaving your body is abrupt in 46.9 percent of cases but can also vary from slow (21.9 percent) to gradual and very slow (15.6 percent).[26] Many OBEs are short, and one recent study found a duration of less than five minutes in nearly 40 percent of cases and less than half a minute in almost 10 percent. In a little more than half the cases, the subjects "see" their body from an external perspective, and 62 percent do so from a short distance only.[27] Many OBEs involve only a passive sense of floating in a body image, though the sense of selfhood is robust. In a recent study more than half the subjects reported being unable to control their movements, whereas nearly a third could. Others experienced no motion at all.[28] Depending on the study, 31 to 84 percent of subjects find themselves located in a second body (but this may also be an indefinite spatial volume), and about 31 percent of OBEs are actually "asomatic"—they are experienced as bodiless and include an externalized visuospatial perspective only. Vision is the dominant sensory modality in 68.8 percent, hearing in 15.5 percent. An older study found the content of the visual scene to be realistic (i.e., not supernatural) in more than 80 percent of cases.[29]

I have always believed that OBEs are important for any solid, empirically grounded theory of self-consciousness. But I had given up on them long ago; there was just too little substantial research, not enough progress over decades, and most of the books on OBEs merely seemed to push metaphysical agendas and ideologies. This changed in 2002, when Olaf Blanke and his colleagues, while doing clinical work at the Laboratory of Presurgical Epilepsy Evaluation of the University Hospital of Geneva, repeatedly induced OBEs and similar experiences by electrically stimulating the brain of a patient with drug-resistant epilepsy, a forty-three-year-old woman who had been suffering from seizures for eleven years. Because it was not possible to find any lesions using neuroimaging methods, invasive monitoring had to be undertaken to locate the seizure focus precisely. During the stimulation of the brain's right angular gyrus, the patient suddenly reported something strongly resembling an OBE. The epileptic focus was located more than 5 cm from the stimulation site in the medial temporal lobe. Electrical stimulation of this site did not induce OBEs, and OBEs were also not part of the patient's habitual seizures.

Initial stimulations induced feelings that the patient described as "sinking into the bed" or "falling from a height." Increasing the current amplitude to 3.5 milliamperes led her to report, "I see myself lying in bed, from above, but I see only my legs and lower trunk." Further stimulations also induced an instantaneous feeling of "lightness" and of "floating" about six feet above the bed. Often she felt as though she were just below the ceiling and legless.

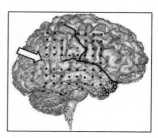

Figure 8 shows the electrode site on the right angular gyrus, where electrical stimulation repeatedly induced not only OBEs but also the feeling of transformed arms and legs or whole-body displacements. Reprinted by permission from Macmillan Publishers Ltd: *Nature*, Volume 419, 19, September 2002.

Meanwhile, not only OBEs but also the "feeling of a presence" have been caused by direct electrical brain stimulation (see figure 9).

Blanke's first tentative hypothesis was that out-of-body experiences, at least in these cases, resulted from a failure to integrate complex

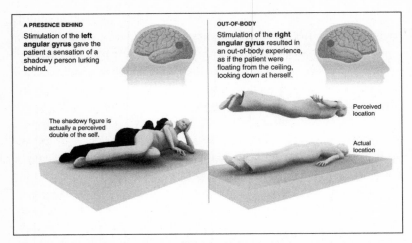

Figure 9: A recent study conducted by Dr. Olaf Blanke provides new scientific insight into experiences more often left to paranormal explanations. Stimulating a part of the brain called the angular gyrus on opposing sides yielded two distinct results: the feeling of a bodily presence behind oneself and an OBE. (Source: Dr. Olaf Blanke. Figure from Graham Roberts/ *The New York Times.*)

somatosensory and vestibular information.[30] In more recent studies, he and his colleagues localized the relevant brain lesion or dysfunction at the temporo-parietal junction (TPJ).[31] They argue that two separate pathological conditions may have to come together to cause an OBE. The first is disintegration on the level of the self-model, brought about by a failure to bind proprioceptive, tactile, and visual information about one's body. The second is conflict between external, visual space and the internal frame of reference created by vestibular information, i.e., our sense of balance. We all move within an internal frame of reference created by our vestibular organs. In vertigo or dizziness, for example, we have problems with vestibular information while experiencing the dominant external, visual space. If the spatial frame of reference created by our sense of balance and the one created by vision come apart, the result could well be the conscious experience of seeing one's body in a position that does not coincide with its felt position.

It is now conceivable that some OBEs could be caused by a cerebral dysfunction at the TPJ. In epileptic patients who report experiencing OBEs, a significant activation at the TPJ can be observed when electrodes are implanted in the left hemisphere.[32] Interestingly, when healthy subjects are asked to imagine their bodies being in a certain position, as if they were seeing themselves from a characteristic perspective of the OBE, same brain region is activated in less than half a second. If this brain region is inhibited by a procedure called transcranial magnetic stimulation, this transformation of the mental model of one's body is impaired. Finally, when an epileptic patient whose OBEs were caused by damage to the temporo-parietal junction was asked to simulate mentally an OBE self-model, this led to a partial activation of the seizure focus. Taken together, these observations point to an anatomical link among three different but highly similar types of conscious experiences: real, seizure-caused OBEs; intended mental simulations of OBEs in healthy subjects; and intended mental simulations of OBEs in epileptic patients.

Recent findings show that the phenomenal experience of disembodiment depends not just on the right half of the temporo-parietal junction but also on an area in the left half, called the extrastriate body-area. A number of different brain regions may actually contribute to the experience. Indeed, the OBE may turn out not to be one single and unified target phenomenon. For example, the phenomenology of exiting the body varies greatly across different types of reports: The initial seconds clearly seem to differ between spontaneous OBEs in healthy subjects and those experienced by the clinical population, such as epileptic patients. The

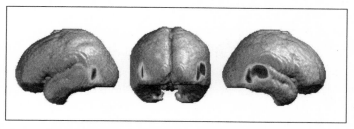

Figure 10: Brain areas that are active in mental transformations of one's body, predominantly at the right temporo-parietal junction. (Figure courtesy of Olaf Blanke, from Blanke et al., "Linking Out-of-Body Experience and Self-Processing to Mental Own-Body Imagery and the Temporoparietal Junction," *Jour. Neurosci.* 25:550–557, 2005.)

onset may also be different in followers of certain spiritual practices. Moreover, there could be a considerable neurophenomenological overlap between lucid dreams (see chapter 5) and OBEs as well as body illusions in general.

VIRTUAL OUT-OF-BODY EXPERIENCES

In 2005, Olaf, his PhD student Bigna Lenggenhager, and I embarked on a series of virtual-reality experiments. Our first goal was to turn the OBE into a fully replicable phenomenon in healthy subjects. Proper research required that we be able to investigate and repeat out-of-body experiences in the lab. The guiding question was whether there could be an integrated kind of bodily self-consciousness that is a phenomenal confabulation. In short, could one experience a hallucinated and a bodily self at the same time, a full-body analog of the rubber-hand illusion?

Here is an example of one of our early experimental protocols using the paraphernalia of virtual reality: a head-mounted display (HMD) consisting of goggles that showed two separate images to each eye, creating the three-dimensional illusion of being in a virtual room. Subjects were able to see their own backs, which were filmed from a distance of 2 meters and projected into the three-dimensional space in front of them with the help of a 3D-encoder. When I acted as the subject of the experiment, I felt as if I had been transposed into a 3D-version of René Magritte's painting *La reproduction interdite*. Suddenly I saw myself from the back, standing in front of me.

Figure 11: Magritte's *La Reproduction Interdite* (1937)

While I was looking at my own back as seen in the head-mounted display, Bigna Lenggenhager was stroking my back, while the camera was recording this action. As I watched my

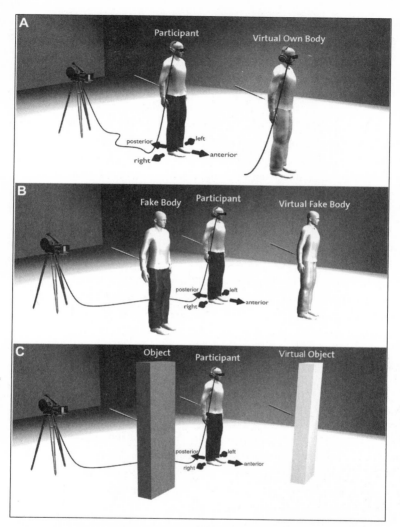

Figure 12: Creating a whole-body analog of the rubber-hand illusion. (A) Participant (darker trousers) sees through a HMD his own virtual body (lighter-colored trousers) in 3D, standing 2 m in front of him and being stroked synchronously or asynchronously at the participant's back. In other conditions, the participant sees either (B) a virtual fake body (lighter trousers) or (C) a virtual noncorporeal object (light gray) being stroked synchronously or asynchronously at the back. Dark colors indicate the actual location of the physical body or object, whereas light colors represent the virtual body or object seen on the HMD. Illustration by M. Boyer.

own back being stroked, I immediately had an awkward feeling: I felt subtly drawn toward my virtual body in front of me, and I tried to "slip into" it. This was as far as things went.

Our studies became more systematic. All of our subjects would be shown their own backs being stroked (this was the "own-body condition") and in a subsequent test would be shown either the back of a mannequin (the "fake-body condition") or a large rectangular slab (which didn't look like a body at all, the "object condition") being stroked. An additional condition was the degree of synchronicity between the seen and the felt stroking, which could be varied by projecting the camera image into cyberspace with a certain time lag.

Afterwards, an independent measure for the strength of the illusion was introduced. They were blindfolded, moved around and disoriented, as in a game of blindman's buff, and then asked to return to their initial position.

At the end of the experiment, the subjects were asked to fill out a questionnaire about their experiences. Results showed that for the synchronous conditions in which they were observing either their own body or a mannequin, they often felt as though the virtual figure was their own body, actually identifying with and "jumping into" it. This impression was less likely to occur in the case of the wooden slab, as well as in all of the asynchronous conditions. The synchronous experiments also showed a significantly larger shift by the subjects toward the projected real or fake body than did the asynchronous control conditions. In other control conditions, subjects observed a screen without a body in it and were then displaced (visual scene), or were simply displaced only. These data suggest that locating the "self" in the case of conflicting visual and somatosensory input is as prone to error as was reported for a body part in the rubber-hand illusion.

Here is what I call the "embedding principle": The bodily self is phenomenally represented as inhabiting a volume in space, whereas the seeing self is an extensionless point—namely, the center of projection for our visuospatial perspective, the geometrical origin of our perspectival visual model of reality. Normally this point of origin (behind the eyes, as if a little person were looking out of them as one looks out a window) is within the volume defined by the felt bodily self. Yet, as our experiments demonstrated, seeing and bodily self can be separated, and the fundamental sense of selfhood is found at the location of the visual body representation.

THE ESSENCE OF SELFHOOD

Why is all this information important for the philosophy of the conscious self? Can it really help us to find the conceptual *essence* of selfhood, to pinpoint what all self-conscious beings in the universe have in common? Is it really a step toward the big picture mentioned in the Introduction? The answer is yes: What we really want are the constitutive conditions for selfhood. We want to know what is truly necessary and what is perhaps only sufficient to bring about an Ego, the fundamental feeling of "being someone." For example, in our quest for the core of the conscious self, it would be progress if one could differentiate between what is merely causally enabling, and what is strictly necessary under the laws of nature holding in this universe. Our experiments demonstrate that agency is not necessary, because they selectively manipulate only two dimensions: self-*identification* (with the content of a conscious body image) and self-*localization* (in a spatial frame of reference). They do so with the subject in a passive condition, without will or bodily agency. This shows how the target phenomenon—self-consciousness—can be causally controlled by multisensory conflict alone. That is important because if we combine the discovery that this can be achieved simply by creating a conflict between sight and touch with the fact that the shift in visual perspective during OBEs can also be caused by an epileptic seizure or by direct stimulation with an electrode in the brain, we get a much better idea of what the simplest form of self-consciousness might be. It must be something very local, something in the brain itself, and it is independent of motor control, of moving your body.[33]

We know more: A seeing self also is not necessary. You can shut the windows in front of the little man behind your eyes by closing your eyelids. The seeing self disappears; the Ego remains. You can be a robust, conscious self even if you are emotionally flat, if you do not engage in acts of will, and also in the absence of thought. Emotions, will, and thoughts are not necessary to the fundamental sense of selfhood. Every meditator (remember chapter 1) can confirm that you may settle into a calm, emotionally neutral state, deeply relaxed and widely alert, a state of pure observation, without any thought, while a certain elementary form of bodily self-consciousness remains. Let us call this "selfhood-as-embodiment."

So what is the essence? Location in space and time plus a transparent body image seem to be very close. The rubber-hand illusion manipulates only the experience of ownership of body parts. The full-body illusion manipulates ownership of the body *as a whole*. Could this finally be the simplest form of selfhood, something we could metaphorically describe as the fundamental experience of "global ownership"? This, I believe, is a misleading idea. Global ownership is a dangerous concept, because it introduces two distinct entities plus a relation, the body and an invisible self, someone who possesses the body. It is the body that possesses itself: Owning something means to be able to control it, and selfhood is intimately related to the very moment in which the body discovers that it can control itself—as a whole. It is exactly what happens when you wake up in the morning, when you "come to yourself."

Here is an interim theory: Minimal self-consciousness is not control, but what makes control *possible*. It includes an image of the body in time and space (location) plus the fact that the organism creating this image does not recognize it *as* an image (identification). So we must have a Now, plus a spatial frame of reference, and a transparent body-model. Then we need a visual (or auditory) perspective originating within the body volume, a center of projection embedded in the volume of the body. But the really interesting step is the one from the minimal self to a slightly more robust *first-person* perspective. It is the step from selfhood-as-embodiment to selfhood-*as-subjectivity*.

The decisive transition takes place when the system is already given to itself through minimal self-consciousness and then, in addition, represents itself as being *directed* toward an object. I believe this happens exactly when we first discover that we can control the focus of attention. We understand that we can draw things from the fringe of consciousness into the center of experience, holding them in the spotlight of attention or deliberately ignoring them—that we can actively control *what* information appears in our mind. Now we have a perspective, because we have an inner image of ourselves as actually representing, as subjects directed at the world. Now we can, for the first time, also attend to our own body as a whole—we become self-directed. Inwardness appears. The essence of this slightly stronger form of selfhood—what a philosopher might call its "representational content"—is attentional agency plus the realization that the body is now available for global control. It is inner knowledge, not of ongoing motor behavior or of perceptual and attentional processing directed at the world or single body parts, but of the body as a single multisensory whole, which now becomes functionally *available* for global control. Conscious selfhood is a deep-seated form of knowledge about oneself, providing information about new causal properties. This inner knowledge has nothing to do with language or concepts. An animal could have it.

What exactly is this "coming to"? Here is another lesson to be learned from the careful study of OBEs: Some OBErs act, but others have a passive experience of floating in a body image; often the second body is not even available for conscious control, yet the sense of selfhood is robust. In a recent study, 53.1 percent of subjects reported not being able to control their own movements (whereas 28.1 percent did, and others didn't experience motion at all).[34] So it clearly is the more subtle experience of controlling the focus of attention, which seems to be at the heart of inwardness—selfhood-as-subjectivity is intimately related to "modeling mental resource allocation" as some sober computational neuroscientist might say. The correct philosophical term would be "epistemic control": The mental action of expanding your knowledge about the world, for example, by selecting *what* you will know, while at the same time excluding what you will, for now, ignore. What this adds is a strong first-person perspective, the experience of being *directed* at an object. Subjective awareness in this sense of having a perspective by being directed at the world is body image (in space and time) plus the experience of attentional control; inwardness appears when we attend to the body itself. Recall how, in chapter 2, I said that consciousness is the space of attentional agency. Selfhood as inwardness emerges when an organism for the first time actively attends to its body as a whole. If a global model of the body is integrated into the space of attentional agency, a richer phenomenal self emerges. It is not necessary to think, it is not necessary to move; the availability of the body as a whole for focal attention is enough to create the most fundamental sense of selfhood-as-inwardness—that is, the ability to become actively *self-directed* in attention. The body model now becomes a self-model in a philosophically more interesting sense: The organism is now potentially directed at the world and at itself at the same time. It is the body *as subject.*

But again—who controls the focus of attention? In our *Video Ergo Sum* study, who is the entity misidentifying itself? Might we nevertheless have a soul, or some sort of astral body, that could survive even bodily death and experience some kind of virtual reincarnation? Will we soon achieve artificial immortality by entering into software worlds designed by human beings, through advanced Magritte-style "forbidden reproduction," deliberately identifying ourselves with virtual bodies and virtual persons we have created for ourselves?[35] Is the phenomenal world itself, perhaps, just virtual reality?

WE LIVE IN A VIRTUAL WORLD

The history of philosophy has shown that technological metaphors have considerable limitations; nevertheless, virtual reality is a useful one. Nature's virtual reality is conscious experience—a real-time world-model that can be viewed as a permanently running online simulation, allowing organisms to act and interact.

Millions of years ago, nature's virtual reality achieved what today's software engineers still strive for: the phenomenal properties of "presence" and "full immersion." From an engineering point of view, the problems involved in creating successful virtual environments are problems of advanced interface design. A virtual interface is a system of transducers, signal processors, hardware, and software. It creates an interactive medium that conveys information to the user's senses while constantly monitoring the user's behavior and employing it to update and manipulate the virtual environment.

Conscious experience, too, is an interface, an invisible, perfect internal medium allowing an organism to interact flexibly with itself. It is a control device. It functions by creating an internal user interface—an "as if" (that is, virtual) reality. It filters information, has a high bandwidth, is unambiguous and reliable, and generates a sense of presence. More important, it also generates a sense of self. The self-model is much like the mouse pointer on the virtual desktop of your PC—or the little red arrow on the subway map that advises "You are here." It places you at the center of a behavioral space, of your consciously experienced world-model, your inner virtual reality.

The Ego is a special part of this virtual reality. By generating an internal image of the organism as a whole, it allows the organism to appropriate its own hardware. It is evolution's answer to the need for explaining one's inner and outer actions to oneself, predicting one's behavior, and monitoring critical system properties. Finally, it allows the system to depict internally the history of its actions as its *own* history. (Autobiographical memory, of course, is one of the most important layers of the human self-model, enabling us to appropriate our own history, inside-time and outside-time, the Now and the past.) Consciousness gives you flexibility, and global control gives you the Ego. On the level of conscious experience, this process of functionally appropriating one's hardware—one's body—in a holistic fashion is mirrored as the sense of global ownership, or minimal selfhood.

Nature, it seems, was engaged in advanced interface design long before we were. It is interesting to note that the best theorists researching virtual environments today not only employ philosophical notions such as "presence" or "situatedness" but also talk about the "virtual body."[36] For them, a virtual body is part of an extended virtual environment. It is a tool that functions much like the little red arrow or the mouse pointer. If the virtual body is employed as an interface, it can even be used to control a robot at a distance. The related concept of a "slave robot" is particularly interesting. To achieve such *telepresence*, there must be a high correlation between the human operator's movements and the actions of the slave robot. (Recall the monkey controlling the robot arm? Now monkeys can even remote-control the real-time walking patterns of humanoid robots halfway around the world, from Duke University in America to the Computational Brain Project of the Japan Science and Technology Agency in

Japan, and through a recording of their brain activity only. As Professor Miguel Nicolelis reports, "The most stunning finding is that when we stopped the treadmill and the monkey ceased to move its legs, it was able to sustain the locomotion of the robot for a few minutes—just by thinking—using only the visual feedback of the robot in Japan.")[37]

Ideally, a human operator would identify his or her own body with that of the slave robot, achieving this with the help of the virtual body, which functions as an interface. Again, nature did just that millions of years ago: Like a virtual body, the phenomenal self-model is an advanced interface designed to appropriate and control a body. Whereas in the case of the virtual body, the slave robot may be thousands of miles away, in the case of the Ego, the target system and the simulating system are identical: The conscious experience of being a subject arises when a single organism learns to *enslave itself.*

The emergence of an Ego Tunnel created a much more efficient way of controlling one's body. Controlling one's body meant controlling one's behavior and one's perceptual machinery. But it also meant directing one's thoughts and regulating one's emotional states. The integrated conscious self-model is a special region of the high-dimensional user interface that emerged in our brains. It is a particularly user-friendly interface, allowing a biological organism to direct its attention to a critical subset of its own global properties. Having a self-model is like adaptive user-modeling, except that it is self-directed and taking place internally. In an important sense, the resulting Ego is a fiction; however, it is also a wonderfully efficient control device. You could also say that it is an entirely new window on reality.

I claim that phenomenal first-person experience and the emergence of a conscious self are complex forms of virtual reality. A virtual reality is a *possible* reality. As anyone who has worn a head-mounted display or played modern video games knows, we can sometimes forget the "as if" completely—the possible can be experienced as the real. In a way, the conscious parts of our brains are like the body's head-mounted display: They immerse the organisms in a simulated behavioral space.

Together, the embodied brain and the PSM, the phenomenal self-model, work much like a total flight simulator. Before we get to the little word "total," let's look at why a flight simulator is a good metaphor for the way our consciousness works. A flight simulator is, of course, a training device that helps pilots learn to control an airplane successfully. To achieve this, the simulation must be as accurate as possible in integrating two different sources of sensory information: vision and the proprioceptive sense of balance. During a simulated takeoff, for example, the pilot not only has to see the runway, but he also has to feel the acceleration of the "as if" plane—and in relation to his own body.

Advanced flight simulators have replaced the movable cockpit mockup and computer screen with a head-mounted display; two slightly displaced monitors create three-dimensional surround graphics. A special programming technique called infinity optics allows the pilot to look at remote objects "through the windows" of the cockpit, even though the computer-generated images are only a few inches from his face. The mock-up's movable platform has been replaced with seat shakers that simulate a range of realistic kinesthetic sensations, such as acceleration or turbulence. Moreover, so the pilot can learn to use on-board instruments and get to know how the aircraft will react to different operations, the simulations of visual and kinesthetic input are constantly updated at great speed and with maximum accuracy.

The human brain can be compared to a modern flight simulator in several respects. Like a flight simulator, it constructs and continuously updates an internal model of external reality by using a continuous stream of input supplied by the sensory organs and employing past experience as a filter. It integrates sensory-input channels into a global model of reality, and it does so in real time. However, there is a difference. The global model of reality constructed by our brain is updated at such great speed and with such reliability that we generally do not experience it as a model. For us, phenomenal reality is not a simulational space constructed by our brains; in a direct and experientially untranscendable manner, it is the world we live in. Its virtuality is hidden, whereas a flight simulator is easily recognized as a flight simulator—its images always seem artificial. This is so because our brains continuously supply us with a much better reference model of the world than does the computer controlling the flight simulator. The images generated by our visual cortex are updated much faster and more accurately than the images appearing in a head-mounted display. The same is true for our proprioceptive and kinesthetic perceptions; the movements generated by a seat shaker can never be as accurate and as rich in detail as our own sensory perceptions.

Finally, the brain also differs from a flight simulator in that there is no user, no pilot who controls it. The brain is like a *total flight simulator,* a self-modeling airplane that, rather than being flown by a pilot, generates a complex internal image of itself within its own internal flight simulator. The image is transparent and thus cannot be recognized as an image by the system. Operating under the condition of a naive-realistic self-misunderstanding, the system interprets the control element in this image as a nonphysical object: The "pilot" is born into a virtual reality with no opportunity to discover this fact. The pilot is the Ego. The total flight simulator generates an Ego Tunnel but is completely lost in it.

If the virtual self functions extremely well, the organism using it is completely unaware of its "as if" nature. The self-model activated in the human brain has been optimized over millions of years. The process that constructs it is fast, reliable, and has a much higher resolution than any of today's virtual-reality games. As a result, the virtuality of the phenomenal self-model tends to be invisible to its user. But strictly speaking, it is simply the best hypothesis the system has about its own current state—presented in a new, highly integrated data format. To illustrate this point, let's look at a classic experiment in modern neuropsychology.

PHANTOM LIMBS

Following amputation, many patients experience a so-called phantom limb at some point—the persistent and unmistakable impression that the lost limb is still present, still part of their body.[38] These phantom limbs feel somewhat less real than the rest of the body, a bit "ghostly." Silas Weir Mitchell, the American neurologist who introduced the concept of phantom limbs in 1871, spoke of "ghostly members" haunting people like "unseen ghosts of the lost part."[39] Often, the phantom recedes gradually and finally disappears; in some cases, however, phantom limbs can persist for months or even years. Patients often have painful sensations in their phantom limb. Sometimes, as in the now-classic case soon to be described, the phantom is "paralyzed," creating the subjective impression that the absent limb is frozen in a fixed position and cannot be moved.

In a set of experiments involving a patient with a paralyzed phantom limb, V. S. Ramachandran and his UCSD colleagues demonstrated the virtuality of the bodily self-model.[40] They constructed a "virtual-reality box" to show to what extent the content of the self-model depends on perceptual-context information. Their idea was that by manipulating the perceptual-context information—which in turn constrains the information-processing activity in the brain—the content of the bodily self-model can be changed.

Their virtual-reality box was quite simple. A mirror was placed vertically inside a cardboard box open at the top, and two holes were cut in the front of the box, to either side of the mirror. The experimenters asked Philip, a patient who had been suffering from a paralyzed phantom limb for many years, to insert both of his arms—that is, his right arm and his left "phantom arm"—through the holes in the box. Then he was told to observe the reflection of his real hand in the mirror. The mirror image of his right hand was used to create the visual illusion that he did actually have two hands. Next, he was asked to make symmetrical movements with both his real arm and his phantom arm.

What would happen to the content of Philip's self-model if the imagined movements of his phantom arm were simultaneously matched with visual input? What would happen to his paralyzed phantom if he could see the movements of a hand in the mirror? Ramachandran described the outcome:

> I asked Philip to place his right hand on the right side of the mirror in the box and imagine that his left hand (the phantom) was on the left side. "I want you to move your right and left arm simultaneously," I instructed.
>
> "Oh, I can't do that," said Philip. "I can move my right arm, but my left arm is frozen. Every morning when I get up, I try to

move my phantom because it's in this funny position and I feel
that moving it might help relieve the pain. But," he said, looking
down at his invisible arm, "I never have been able to generate a
flicker of movement in it."

"Okay, Philip, but try anyway."

Philip rotated his body, shifting his shoulder, to "insert" his
lifeless phantom into the box. Then he put his right hand on the
other side of the mirror and attempted to make synchronous
movements. As he gazed into the mirror, he gasped and then
cried out, "Oh, my God! Oh, my God, doctor! This is unbeliev-
able. It's mind-boggling!" He was jumping up and down like a
kid. "My left arm is plugged in again. It's as if I'm in the past. All
these memories from years ago are flooding back into my mind.
I can move my arm again. I can feel my elbow moving, my wrist
moving. It's all moving again."

After he calmed down a little I said, "Okay, Philip, now close
your eyes."

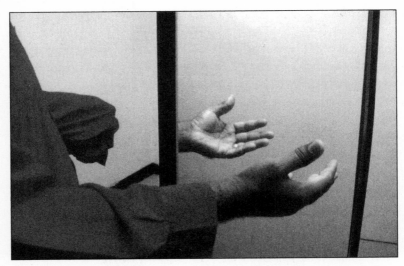

Figure 13: Mirror-induced synesthesia. Making part of a hallucinated self available for
conscious action control by installing a virtual source of visual feedback. Picture
courtesy of V. S. Ramachandran.

"Oh, my," he said, clearly disappointed. "It's frozen again. I feel my right hand moving, but there's no movement in the phantom."

"Open your eyes."

"Oh, yes. Now it's moving again."[41]

The phantom movement in this experiment is the content of the conscious self-model. In the real world, there is no limb that can be felt or controlled. In his moving phantom limb, Philip experiences—and controls—a part of his bodily self that does not exist. Just as in the rubber-hand illusion, the experiential property of ownership seamlessly spreads into the hallucinated part of the bodily self: The moving phantom limb is owned, just as the rubber hand is owned. On the intellectual level, Philip understands perfectly well that the phantom limb does not exist. (This fact is *cognitively available* to him, as a philosopher might say.) But the subjective experience of his phantom arm actually moving is robust and realistic. And, as opposed to the rubber-hand illusion, there is an additional quality—namely, the phenomenal experience of *agency*. A full-blown bodily Ego is in place.

In order to survive, biological organisms must not only successfully predict what is going to happen next in their immediate environment but also be able to predict accurately their behavior and bodily movements along with their consequences. The self-model is a real-time predictor. This is how our best current theories explain what happened to Philip: In our brains, we have a body emulator that uses motor commands to predict the likely proprioceptive and kinesthetic feedback that results from moving our limbs in a certain way. For our actions to be successfully controlled, we cannot wait for the actual feedback from our arms and legs as we move through the world. We need an internal image of our body as a whole that predicts the likely consequences of, say, an attempt to move our left arm in a certain way. In order to be really efficient, we need to know in advance what this would feel like. Furthermore, by "taking it offline," we can use our body emulator to create motor imagery in our mind—to plan or imagine our body movements without actually executing them.

This body emulator, which constantly generates forward simulations, is a fundamental part of the human self-model and the centerpiece of the Ego Tunnel. Philip's self-model had learned that whatever motor commands he issued to his amputated arm, there would be no feedback telling him about a changed limb position. To be sure, the image of his arm was still there, burned into his brain. It had adapted to zero feedback and was therefore frozen. Ramachandran's ingenious idea was to use a mirror as a source of virtual information, allowing the virtual emulator to perform a successful update. When Philip tried to move both his real arm and his phantom arm, the changes in the visual self-model perfectly matched the motor commands fed into the body-state predictor in Philip's brain. The conscious experience that his missing left arm was actually moving and under volitional control followed suit.

Now we can understand why our self-model is a virtual model. Clearly, Philip's moving left arm is just a simulation. It is an "as if" arm;

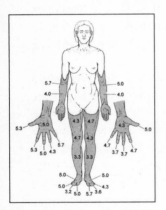

what has turned into a new possibility for the brain is portrayed to Philip as a reality. If one does not think about it but simply attends to the experience itself, the moving phantom limb can perhaps be experienced as just as realistic as the rest of the body; they are both parts of the same unified self, and they are both under volitional control. But exactly *how* real the parts of our self-model appear depends on many different factors.

One interesting fact about phantom-limb experiences is that they also happen to people who were born missing certain limbs. A recent case study conducted by Swiss neuroscientist Peter Brugger and his colleagues of the University Hospital in Zürich used a seven-point scale to rate the subjectively perceived vividness of phantom limbs.[42] Interestingly, the ratings showed highly consistent judgments across sessions for their subject, "AZ," a forty-four-year-

Figure 14: Evidence for an innate component of the body model? Phantoms (shaded areas) in a subject born without limbs. The numbers are vividness ratings for the felt presence of different phantom body parts on a seven-point scale from 0 (no awareness) to 6 (most vivid impression). Picture courtesy of Peter Brugger, Zürich.

old university-educated woman born without forearms and legs. For as long as she could remember, she had experienced mental images of her nonexistent forearms (including fingers) and legs (including feet and the first and fifth toes). But, as the figure shows, these phantoms were not as realistic as the content of her nonhallucinatory body model.

Moreover, she reported that "[a]wareness of her phantom limbs is transiently disrupted only when some object or person invades their felt position or when she sees herself in a mirror." Functional magnetic resonance imaging (fMRI) of her imagined phantom hand movements showed no activation of primary sensorimotor areas, but did show activity in the bilateral premotor and parietal cortex. Transcranial magnetic stimulation of the sensorimotor cortex consistently elicited sensations in the phantom fingers and hand, on the side opposite the stimulation. Premotor and parietal stimulation evoked similar phantom sensations, though without showing motor-evoked potentials in the stump. Brugger's data demonstrate that body parts that never physically developed can be represented in sensory and motor areas of the cortex.

The fascinating question is this: Are AZ's hallucinated forearms and legs components of an *innate* body model—perhaps of a nucleus that continues developing after birth? Or could they have been "mirrored into" the patient's self-model through the visual observation of the movements of other human beings (see chapter 6 on the Empathic Ego)? What exactly is it that *you* feel as your own body, right now, as you are reading these words? At this point in our investigations into consciousness, it seems obvious that we are never in direct contact with our physical bodies but rather with a particular kind of representational content. But what exactly is it that is represented in this layer of our conscious self? In the second book of his famous work *De anima*, Aristotle said that the soul is simply the *form* of the body and that it perishes at death. Is that what we have newly rediscovered by studying phantom limbs, the "inner form" of the body and the global model of its shape? Spinoza said the soul is the idea that the body develops of itself because "the object of our soul is the body as it exists, and nothing else." [43] Again, it is intriguing to see how classical philosophical ideas contribute to a deeper understanding of what it means to be an embodied self.

Ramachandran's and Brugger's experiments demonstrate that the experiential content of the bodily self-model is the content of an ongoing simulation, part of a dynamic control mechanism. At any given time, the content of bodily experience is the best hypothesis that the system has about its current body state. The brain's job is to simulate the body *for* the body and to predict the consequences of the body's movements, and the instrument it uses is the self-model. This process takes place in the real world, so it is time-consuming and necessarily generates a lag between the actual state of the body and the self-model's content.

Normally we're unaware of this process, because nature engineered it so efficiently that errors rarely occur. But the simple fact remains: You are never in direct contact with your own body. What you feel in the rubber-hand illusion, what AZ feels, or what Philip feels when his left arm is "plugged in" is exactly the same as what you feel when you attend to the sensation of your hands holding this book right now or to the feeling of pressure and resistance when you lean back in your chair. What you experience is not reality but virtual reality, a possibility. Strictly speaking, and on the level of conscious experience alone, you live your life in a virtual body and not in a real one. This point will become clearer when we consider "offline states" in the chapter on dreaming and lucid dreaming. But first, let us have a look at another essential feature of phenomenal selfhood—the transition from ownership to agency.

FOUR

FROM OWNERSHIP TO AGENCY TO FREE WILL

Before the use of external tools could develop, a neurodynamic tool had to be in place in our brains. I have been calling this inner tool the PSM, the phenomenal self-model, a distinct and coherent pattern of neural activity that allows you to integrate parts of the world into an inner image of yourself as a whole. Only if you have a self-model can you experience your hands and your arms as parts of your *own* body. Only if you have a self-model can you experience certain cognitive processes in your brain as your *own* thoughts and certain events in the motor parts of your brain as your *own* intentions and acts of will. Our next step is the step from ownership to agency.

THE ALIEN HAND

Imagine that about ten days after undergoing heart surgery you notice a weakness in your left side and experience difficulties walking. For the past three days, you have also had a more specific problem: Somehow, you keep losing control of your left hand—it is acting on its own. Last night, you awoke several times because your left hand was trying to choke you, and you had to use your right hand to fight it off. During the day, your left hand sometimes unbuttons your hospital gown just after your right hand has buttoned it up. Your left hand crushes the paper cups on your tray or starts fighting with your right hand while you're trying to answer the phone. It's an unpleasant situation, to say the least—as if someone "from the moon" were controlling your hand. Sometimes you wonder whether it has a mind of its own.[1]

What does it mean for something to "have a mind of its own"? Having a mind means possessing inner states that have content and embedding such thoughts and inner images of the world into a self-model. Then the organism harboring them can know that they are occurring within itself. So far, so good. But there is an important aspect of having a mind of your own that we've not yet discussed: You also need explicit representations of *goal-states*—your requirements, your desires, your values, what you want to achieve by acting in the world. And you need a conscious Ego to appropriate these goal-states, to make them your own. Philosophers call this having "practical intentionality": Mental states are often directed at the fulfillment of your personal goals. Having a mind means being not only a thinker and a knower but also an *agent*—an acting self, with a will of one's own.

That is where the Alien Hand syndrome, the neurological disorder just described, comes in. The syndrome was first described in 1908, but the term was not introduced until 1972, and it still isn't clear what the necessary and sufficient conditions in the brain for this kind of disorder are.[2] The alien hand crushing cups on the tray and fighting with the healthy right hand seems to have a will of its own. When the alien hand begins unbuttoning the patient's gown, this is not automatic behavior like the knee-jerk reflex; it appears to be guided by an explicit goal-representation. Apparently a little agent is embedded in the bigger agent—a subpersonal entity pursuing its own goals by hijacking a body part that belongs to the patient. In another typical case, a patient will pick up a pencil and begin scribbling with one hand, reacting with dismay when she becomes aware of this. She will then immediately withdraw the pencil, pull the alien hand to her side with her "good" hand, and indicate that she did not initiate the scribbling herself.[3] Another such case study describes the patient's left hand groping for nearby objects and picking and pulling at her clothes to the point that she refers to her errant hand as an autonomous entity.[4]

These cases are interesting from a philosophical point of view, because any convincing philosophical theory of the conscious self will have to explain the dissociation of ownership and agency. Patients suffering from Alien Hand syndrome still experience the hand as their own hand; the conscious sense of ownership is still there, but there is no corresponding experience of *will* in the patient's mind. As philosophers say, the "volitional act" is missing, and the goal-state driving the alien hand's behavior is not represented in the person's conscious mind. The fact that the arm is clearly a subpersonal part of the body makes it even more striking to see how the patient automatically attributes something like intentionality and personhood to it, treating it as an autonomous agent. This conflict between the hand and the *willing* self can even become a conflict between the hand and the *thinking* self. For instance, when one patient's left hand made a move he did not wish to make in a game of checkers, he corrected the move with his right hand. Then, to his frustration, the isolated functional module in his brain that was driving his left arm caused it to repeat the unwanted move.[5]

Here is the philosophical problem: Is the unwanted move in the game of checkers an action—that is, a bodily movement directly caused by an explicit goal-representation—or is it only an event, something that just happens, caused by something else? At one extreme of the philosophical spectrum, we find denial of the freedom of will: No such things as "actions" or "agents" exist, and, strictly speaking, predetermined physical events are all that have ever existed. We are all automata. If our hardware is damaged, individual subsystems may act up—a sad fact, but certainly no mystery. The other extreme is to hold that there are no blind, purely physical events in the universe at all, that every single event is a goal-driven action, caused by a person—for instance, by the mind of God. Nothing happens by chance; everything is purposeful and ultimately willed.

In fact, in some psychiatric syndromes, patients experience every consciously perceived event in their environment as directly caused by themselves. In other mental diseases, such as schizophrenia, one may feel that one's body and thoughts are remote-controlled and that the whole world is one big machine, a soulless and meaningless mechanism grinding away. Note that both types of observations illustrate my claim in chapter 1 that we must view the brain as a reality engine: It is a system that constantly makes assumptions about what exists and what doesn't, thereby creating an inner reality including time, space, and causal relations. Psychiatric diseases are reality-models—alternate ontologies developed to cope with serious and often specific problems. Interestingly, in almost all cases these alternate ontologies can be mapped onto a philosophical ontology—that is, they will correspond to some well-established metaphysical idea about the deeper structure of reality (radical determinism, say, or the omnipotent, omnipresent God's-eye view).

But to return to the original question: Do actions as such really exist? A position between the two philosophical extremes would define "action" as a particular kind of physical event. Most events in the physical universe are only events, but an extremely tiny subset are also actions—that is, events caused by an explicit goal-representation in the conscious mind of a rational agent. Goal-states must be owned by being part of a self-model. No Ego Tunnel, no action.

The alien hand, however, is *not* a distinct entity with an Ego Tunnel. It is just a body part and has no self-model. It does not know about its existence, nor does a world appear to it. Due to a brain lesion, it is driven by one of the many unconscious goal-representations constantly fighting for attention in your brain—plausibly, it is driven by visually perceived objects in your immediate vicinity that give rise to what psychologists and philosophers call *affordances*. There is good evidence that the brain portrays visual objects not only as such but also in terms of possible movements: Is this something I could grasp? Is this something I could unbutton? Is this something I could eat or drink?

The self-model is an important part of the selection mechanism. Right now, as you are reading this book, it is protecting you from these affordances, preventing them from taking over parts of your body. If I were to put a plate of your favorite chocolate cookies in front of you and if you had the firm determination not to reach for it, how long could you keep concentrating on the book? How long before a brief episode of Alien Hand syndrome would pop up and your left hand would do something you hadn't told it to do? The stronger and more stable your self-model, the less susceptible you are to the affordances surrounding you. Autonomy comes in degrees; it has to do with immunization, with shielding yourself from infection by potential goal-states in the environment.

The phenomenal experience of ownership and the phenomenal experience of agency are thus intimately related, and both are important aspects of the conscious sense of self. If you lose control over your actions, your sense of self is greatly diminished. This is also true of inner actions; for example, many schizophrenics feel that not only their bodies but even their thoughts are controlled by alien forces. One of my pet ideas for many years might well turn out to be true—namely, that thinking is a motor process. Could thoughts be models of successfully terminated actions but from a God's-eye view—that is, independent of your own vantage point? Could they be abstract forms of grasping—of holding an object and taking it in, into your self? As I discuss in the chapter on the Empathic Ego, there is solid empirical evidence showing that the hand is represented in Broca's area, a part of our brain that is of recent evolution, distinguishes us from monkeys, and has to do with language comprehension and abstract meaning. The thinking self would then have grown out of the bodily self, by simulating bodily movements in an abstract, mental space. I have been flirting with this idea for a long time, because it would solve Descartes' mind-body problem; it would show how a thinking thing—a *res cogitans*—could have evolved out of an extended thing, a *res extensa*. And this points to a theme running through much of the recent research on agency and the self: In its origin, the Ego is a neurocomputational device for appropriating and controlling the body—first the physical one and then the virtual one.

There is a kind of agency even more subtle than the ability to experience yourself as a coherent acting self and the direct cause of change: This is what I call *attentional agency*. Attentional agency is the experience of being the entity that controls what Edmund Husserl described as *Blickstrahl der Aufmerksamkeit*—the "ray of attention." As an attentional agent, you can initiate a shift in attention and, as it were, direct your inner flashlight at certain targets: a perceptual object, say, or a specific feeling. In many situations, people lose the property of attentional agency, and consequently their sense of self is weakened. Infants cannot control their visual attention; their gaze seems to wander aimlessly from one object to another, because this part of their Ego is not yet consolidated. Another example of consciousness without attentional control is the dream state, and, as I discuss in the next chapter, the Ego of the dream state is indeed very different from that of the waking state. In other cases, too, such as severe drunkenness or senile dementia, you may lose the ability to direct your attention—and, correspondingly, feel that your "self" is falling apart.

Then there is *cognitive agency,* an interesting parallel to what philosophers call the "cognitive subject." The cognitive subject is a thinker of thoughts and can also ascribe this faculty to herself. But often thoughts just drift by, like clouds. Meditators—like the Tibetan monks in chapter 2—strive to diminish their sense of self, letting their thoughts drift by instead of clinging to their content, attentively but effortlessly letting them dissolve. If you had never had the conscious experience of causing your own thoughts, ordering and sustaining them, being attached to their content, you would never have experienced yourself as a thinking self. That part of your self-model would simply have dried up and withered away. In order to have Descartes' experience of the *Cogito*—the robust experience of being a thinking thing, an Ego—you must also have had the experience of deliberately selecting the contents of your mind. This is what the various forms of agency have in common: Agency allows us to *select* things: our next thought, the next perceptual object we want to focus on, our next bodily movement. It is also the experience of *executive consciousness*—not only the experience of initiating change but also of carrying it through and sustaining a more complex action over time. At least this is the way we have described our inner experience for centuries.

A related aspect that bodily agency, attentional agency, and cognitive agency have in common is the subjective sense of effort. Phenomenologically, it is an effort to move your body. It is also an effort to focus your attention. And it certainly is an effort to think in a concentrated, logical fashion. What is the neural correlate of the sense of effort? Imagine we knew this neural correlate (we will soon), and we also had a precise and well-tested mathematical model describing what is common to all three kinds of experiencing a sense of effort. Imagine you are a future mathematician who can understand this description in all of its intricate detail. Now, given this detailed conceptual knowledge, you introspect your own sense of effort, very gently, but with great precision. What would happen? If you were to gently and carefully attend to, say, the sense of effort going along with an act of will, would it still appear as something personal, something that belongs to you?

The Alien Hand syndrome forces us to conclude that what we call *the will* can be outside our self-model as well as inside it. Such goal-directed movements might not even be consciously experienced at all. In a serious neurological disorder called *akinetic mutism*, patients do nothing but lie silently in their beds. They have a sense of ownership of their body as a whole, but although they are awake (and go through the ordinary sleep-wake cycle), they are not agents: They do not act in any way. They do not initiate any thoughts. They do not direct their attention. They do not talk or move.[6] Then there are those cases in which parts of our bodies perform complex goal-directed actions without our having the conscious experience of these being *our* actions or our goals, without a conscious act of will having preceded them—in short, without the experience of being an agent. Another interesting aspect—and the third empirical fact that any philosophy of the conscious self must explain—is how, for instance, schizophrenics sometimes lose the sense of agency and executive consciousness entirely and feel themselves to be remote-controlled puppets.

Many of our best empirical theories suggest that the special sense of self associated with agency has to do both with the conscious experience of having an intention and with the experience of motor feedback. That is, the experience of selecting a certain goal-state must be integrated with the subsequent experience of bodily movement. The self-model achieves just that. It binds the processes by which the mind creates and compares competing alternatives for action with feedback from your bodily movements. This binding turns the experience of movement into the experience of an action. But note, once again, that neither the "mind" nor the self-model is a little man in the head; there is no one doing the creating, the comparing, and the deciding. If the dynamical-systems theory is correct, then all of this is a case of dynamical self-organization in the brain. If for some reason the two core elements—the selection of a specific movement pattern and ongoing motor feedback—cannot be successfully bound, you might experience your bodily movements as uncontrolled and erratic (or as controlled by someone else, as schizophrenics sometimes do). Or you might experience them as willed and goal-directed but not as self-initiated, as in the Alien Hand syndrome.

HALLUCINATING AGENCY

Thus, selfhood is something independent, because one can retain the sense of ownership yet lose the sense of agency. But can one also hallucinate agency? The answer is yes—and, oddly, many consciousness philosophers have long ignored this phenomenon. You can have the robust, conscious experience of having intended an action even if this wasn't the case. By directly stimulating the brain, we can trigger not only the execution of a bodily movement but also the conscious experience of having the urge to perform that movement. We can experimentally induce the conscious experience of will.

Here's an example. Stéphane Kremer and his colleagues at the University Hospital of Strasbourg stimulated a specific brain region (the ventral bank of the anterior cingulate sulcus) in a female patient with medically intractable epileptic seizures, in order to locate the epileptogenic zone before performing surgery. In this case, the stimulation caused rapid eye movements scanning both sides of the visual field. The patient began to search for the nearest object she could grasp, and the arm that was opposite the stimulated side—her left arm—began to wander to the right. She reported a strong "urge to grasp," which she was unable to control. As soon as she saw a potential target object, her left hand moved toward it and seized it. On the level of her conscious experience, the irrepressible urge to grasp the object started and ended with the stimulation of her brain. This much is clear: Whatever else the conscious experience of will may be, it seems to be something that can be turned on and off with the help of a small electrical current from an electrode in the brain.[7]

But there are also ways of elegantly inducing the experience of agency by purely psychological means. In the 1990s at the University of Virginia, psychologists Daniel M. Wegner and Thalia Wheatley investigated the necessary and sufficient conditions for "the experience of conscious will" with the help of an ingenious experiment. In a study they dubbed "I Spy," they led subjects to experience a causal link between a thought and an action, managing to induce the feeling in their subjects that the subjects had willfully performed an action even though the action had in fact been performed by someone else.[8]

Each subject was paired with a confederate, who posed as another subject. They sat at a table across from each other and were asked to place their fingertips on a little square board mounted on a computer mouse, enabling them to move the mouse together, Ouija-board style. On a computer screen visible to both was a photograph from a children's book showing some fifty objects (plastic dinosaurs, cars, swans, and so on).

The real subject and the confederate both wore headphones, and it was explained to them that this was an experiment meant to "investigate people's feelings of intention for acts and how these feelings come and go." They were told to move the mouse around the computer screen for thirty seconds or so while listening to separate audio tracks containing random words—some of which would refer to one or another object on the screen—along with ten-second intervals of music. The words on each track would be different, but the timing of the music would be the same. When they heard the music, they were to stop the mouse on an object after a few seconds and "rate each stop they made for personal intentionality." Unknown to the

Figure 15: Hallucinated agency. How to make subjects think they initiated a movement they never intended. Figure courtesy of Daniel Wegner.

subject, however, the confederate did not hear any words or music at all but instead received instructions from the experimenters to perform particular movements. For four of the twenty or thirty trials, the confederate was told to stop the mouse on a particular object (each time a different one); these forced stops were made to occur within the prescribed musical interval and at various times after the subject had heard the corresponding word over her headphones ("swan," say).[9]

According to the subjects' ratings, there was a general tendency to perceive the forced stops as intended. The ratings were highest when the corresponding word occurred between one and five seconds before the stop. Based on these findings, Wegner and Wheatley suggest that the phenomenal experience of will, or mental causation, is governed by three principles: The principle of *exclusivity* holds that the subject's thought should be the only introspectively available cause of action; the principle of *consistency* holds that the subjective intention should be consistent with the action; and the principle of *priority* holds that the thought should precede the action "in a timely manner."[10]

The social context and the long-term experience of being an agent of course contribute to creating the sense of agency. One might suspect that the sense of agency is only a subjective appearance, a swift reconstruction after the act; still, today's best cognitive neuroscience of the conscious will shows that it is also a *preconstruction*.[11] Experiencing yourself as a willing agent has much to do with, as it were, introspectively peeping into the middle of a long processing chain in your brain. This chain leads from certain preparatory processes that might be described as "assembling a motor command" to the feedback you get from perceiving your movements. Patrick Haggard, of University College London, perhaps the leading researcher in the fascinating and somewhat frightening new field of research into agency and the self, has demonstrated that our conscious awareness of movement is *not* generated by the execution of ready-made motor commands; instead, it is shaped by preparatory processes in the premotor system of the brain. Various experiments show that our awareness of intention is closely related to the specification of which movements we want to make. When the brain simulates alternative possibilities—say, of reaching for a particular object—the conscious experience of intention seems to be directly related to the selection of a specific movement. That is, the awareness of movement is associated not so much with the actual execution as with an earlier brain stage: the process of preparing a movement by assembling different parts of it into a coherent whole—a motor gestalt, as it were.

Haggard points out that the awareness of intention and the awareness of movement are conceptually distinct, but he speculates that they must derive from a single processing stage in the motor pathway. It looks as though our access to the ongoing motor-processing in our brains is extremely restricted; awareness is limited to a very narrow window of premotor activity, an intermediate phase of a longer process. If Haggard is right, then the sense of agency, the conscious experience of *being someone who acts,* results from the process of binding the awareness of intention together with the representation of one's actual movements. This also suggests what subjective awareness of intention is good for: It can detect potential mismatches with events occurring in the world outside the brain.

Whatever the precise technical details turn out to be, we are now beginning to see what the conscious experience of agency is and how to explain its evolutionary function. The conscious experience of will and of agency allows an organism to *own* the subpersonal processes in its brain responsible for the selection of action goals, the construction of specific movement patterns, and the control of feedback from the body. When this sense of agency evolved in human beings, some of the stages in the immensely complex causal network in our brains were raised to the level of global availability. Now we could attend to them, think about them, and possibly even interrupt them. For the first time, we could experience ourselves as beings with goals, and we could use internal representations of these goals to control our bodies. For the first time, we could form an internal image of ourselves as able to fulfill certain needs by choosing an optimal route. Moreover, conceiving of ourselves as autonomous agents enabled us to discover that other beings in our environment probably were agents, too, who had goals of their own. But I must postpone this analysis of the social dimension of the self for a while and turn to a classical problem of philosophy of mind: the free-dom of the will.

HOW FREE ARE WE?

As noted previously, the philosophical spectrum on freedom of the will is a wide one, ranging from outright denial to the claim that all physical events are goal-driven and caused by a divine agent, that nothing hap-pens by chance, that everything is, ultimately, willed. The most beautiful idea, perhaps, is that freedom and determinism can peacefully coexist: If our brains are causally determined in the *right* way, if they make us causally sensitive to moral considerations and rational arguments, then this very fact makes us free. Determinism and free will are compatible. However, I take no position on free will here, because I am interested in two other points. I address the first by asking one simple question: What does ongoing scientific research on the physical underpinnings of ac-tions and of conscious will tell us about this age-old controversy?

Probably most professional philosophers in the field would hold that given your body, the state of your brain, and your specific environment, you could not act differently from the way you're acting now—that your actions are preordained, as it were. Imagine that we could produce a perfect duplicate of you, a functionally identical twin who is an exact copy of your molecular structure. If we were to put your twin in exactly the same situation you're in right now, with exactly the same sensory stimuli impinging on him or her, then initially the twin could not act dif-ferently from the way you're acting. This is a widely shared view: It is, simply, the scientific worldview. The current state of the physical uni-verse always determines the next state of the universe, and your brain is a part of this universe.[12]

The phenomenal Ego, the experiential content of the human self-model, clearly disagrees with the scientific worldview—and with the widely shared opinion that your functionally identical *doppelgänger* could not have acted otherwise. If we take our own phenomenology se-riously, we clearly experience ourselves as beings that *can* initiate new causal chains out of the blue—as beings that *could* have acted otherwise given exactly the same situation. The unsettling point about modern philosophy of mind and the cognitive neuroscience of will, already apparent even at this early stage, is that a final theory may contradict the way we have been subjectively experiencing ourselves for millennia. There will likely be a conflict between the scientific view of the acting self and the phenomenal narrative, the subjective story our brains tell us about what happens when we decide to act.

We now have a theory in hand that explains how subpersonal brain events (for instance, those that specify action goals and assemble suitable motor commands) can become the contents of the conscious self. When certain processing stages are elevated to the level of conscious experience and bound into the self-model active in your brain, they become available for all your mental capacities. Now you experience them as your own thoughts, decisions, or urges to act—as properties that belong to *you,* the person as a whole. It is also clear why these events popping up in the conscious self necessarily appear spontaneous and uncaused. They are the first link in the chain to cross the border from unconscious to conscious brain processes; you have the impression that they appeared in your mind "out of the blue," so to speak. The unconscious precursor is invisible, but the link exists. (Recently, this has been shown for the conscious veto, as when you interrupt an intentional action at the last instant.)[13] But in fact the conscious experience of intention is just a sliver of a complicated process in the brain. And since *this* fact does not appear to us, we have the robust experience of being able to spontaneously initiate causal chains from the mental into the physical realm. This is the appearance of an agent. (Here we also gain a deeper understanding of what it means to say that the self-model is transparent. Often the brain is blind to its own workings, as it were.)

The science of the mind is now beginning to reintroduce those hidden facts forcefully into the Ego Tunnel. There will be a conflict between the biological reality tunnel in our heads and the neuroscientific image of humankind, and many people sense that this image might present a danger to our mental health. I think the irritation and deep sense of resentment surrounding public debates on the freedom of the will have little to do with the actual options on the table. These reactions have to do with the (perfectly sensible) intuition that certain types of answers will not only be emotionally disturbing but ultimately impossible to integrate into our conscious self-models. This is the first point.[14]

A note on the phenomenology of will: It is not as well defined as you might think; color experience, for example, is much crisper. Have you ever tried to observe introspectively what happens when you decide to lift your arm and then the arm lifts? What exactly is the deep, fine-grained structure of cause and effect? Can you really observe how the mental event causes the physical event? Look closely! My prediction is that the closer you look and the more thoroughly you introspect your decision processes, the more you'll realize that conscious intentions are evasive: The harder you look at them, the more they recede into the background. Moreover, we tend to talk about free will as if we all shared a common subjective experience. This is not entirely true: Culture and tradition exert a strong influence on the way we report such experiences. The phenomenology itself may well be shaped by this, because a self-model also is the window connecting our inner lives with the social practice around us. Free will does not exist in our minds alone—it is also a social institution. The assumption that something like free agency exists, and the fact that we treat one another as autonomous agents, are concepts fundamental to our legal system and the rules governing our societies—rules built on the notions of responsibility, accountability, and guilt. These rules are mirrored in the deep structure of our PSM, and this incessant mirroring of rules, this projection of higher-order assumptions about ourselves, created complex social networks. If one day we must tell an entirely different story about what human will is or is not, this will affect our societies in an unprecedented way. For instance, if accountability and responsibility do not really exist, it is meaningless to punish people (as opposed to rehabilitating them) for something they ultimately could not have avoided doing. Retribution would then appear to be a Stone Age concept, something we inherited from animals. When modern neuroscience discovers the sufficient neural correlates for willing, desiring, deliberating, and executing an action, we will be able to cause, amplify, extinguish, and modulate the conscious experience of will by operating on these neural correlates. It will become clear that the actual causes of our actions, desires, and intentions often have very little to do with what the conscious self tells us. From a scientific, third-person perspective, our inner experience of strong autonomy may look increasingly like what it has been all along: an appearance only. At the same time, we will learn to admire the elegance and the robustness with which nature built only those things into the reality tunnel that organisms needed to know, rather than burdening them with a flood of information about the workings of their brains. We will come to see the subjective experience of free will as an ingenious neurocomputational tool. Not only does it create an internal user-interface that allows the organism to control and adapt its behavior, but it is also a necessary condition for social interaction and cultural evolution.

Imagine that we have created a society of robots. They would lack freedom of the will in the traditional sense, because they are causally determined automata. But they would have conscious models of themselves and of other automata in their environment, and these models would let them interact with others and control their own behavior. Imagine that we now add two features to their internal self- and other-person models: first, the erroneous belief that they (and everybody else) are responsible for their own actions; second, an "ideal observer" representing group interests, such as rules of fairness for reciprocal, altruistic interactions. What would this change? Would our robots develop new causal properties just by falsely believing in their own freedom of the will? The answer is yes; moral aggression would become possible, because an entirely new level of competition would emerge—competition about who fulfills the interests of the group best, who gains moral merit, and so on. You could now raise your own social status by accusing others of being immoral or by being an efficient hypocrite. A whole new level of optimizing behavior would emerge. Given the right boundary conditions, the complexity of our experimental robot society would suddenly explode, though its internal coherence would remain. It could now begin to evolve on a new level. The practice of ascribing moral responsibility—even if based on delusional PSMs—would create a decisive, and very real, functional property: Group interests would become more effective in each robot's behavior. The price for egotism would rise. What would happen to our experimental robot society if we then downgraded its members' self-models to the previous version—perhaps by bestowing insight?

A passionate public debate recently took place in Germany on freedom of the will—a failed debate, in my view, because it created more confusion than clarity. Here is the first of the two silliest arguments for the freedom of will: "But I *know* that I am free, because I experience myself as free!" Well, you also experience the world as inhabited by colored objects, and we know that out there in front of your eyes are only wavelength mixtures of various sorts. That something appears to you in conscious experience and in a certain way is not an argument for anything. The second argument goes like this: "But this would have terrible consequences! Therefore, it *cannot* be true." I certainly share that worry (think of the robot society thought experiment), but the truth of a claim must be assessed independently of its psychological or political consequences. This is a point of simple logic and intellectual honesty. But neuroscientists have also added to the confusion—and, interestingly, because they often underestimate the radical nature of their positions. This will be my second point in this section.

Neuroscientists like to speak of "action goals," processes of "motor selection," and the "specification of movements" in the brain. As a philosopher (and with all due respect), I must say that this, too, is conceptual nonsense. If one takes the scientific worldview seriously, no such things as goals exist, and there is nobody who selects or specifies an action. There is no process of "selection" at all; all we really have is dynamical self-organization. Moreover, the information-processing taking place in the human brain is not even a rule-based kind of processing. Ultimately, it follows the laws of physics. The brain is best described as a complex system continuously trying to settle into a stable state, generating order out of chaos.

According to the purely physical background assumptions of science, nothing in the universe possesses an inherent value or is a goal in itself; physical objects and processes are all there is. That seems to be the point of the rigorous reductionist approach—and exactly what beings with self-models like ours cannot bring themselves to believe. Of course, there can be goal *representations* in the brains of biological organisms, but ultimately—if neuroscience is to take its own background assumptions seriously—they refer to nothing. Survival, fitness, well-being, and security as such are not values or goals in the true sense of either word; obviously, only those organisms that internally represented them as goals survived. But the tendency to speak about the "goals" of an organism or a brain makes neuroscientists overlook how strong their very own background assumptions are. We can now begin to see that even hardheaded scientists sometimes underestimate how radical a naturalistic combination of neuroscience and evolutionary theory could be: It could turn us into beings that maximized their overall fitness by beginning to *hallucinate* goals.

I am not claiming that this is the true story, the whole story, or the final story. I am only pointing out what seems to follow from the discoveries of neuroscience and how these discoveries conflict with our conscious self-model. Subpersonal self-organization in the brain simply has nothing to do with what we mean by "selection." Of course, complex and flexible behaviors caused by inner images of "goals" still exist, and we may also continue to call these behaviors "actions." But even if actions, in this sense, continue to be part of the picture, we may learn that *agents* do not—that is, there is no entity *doing* the acting.[15]

The study of phantom limbs helped us understand how parts of our bodies can be portrayed in the phenomenal self-model even if they do not exist or have never existed. Out-of-body experiences and full-body illusions demonstrated how a minimal sense of self and the experience of "global ownership" can emerge. A brief look at the Alien Hand and the neural underpinnings of the willing self gave us an idea of how the feeling of agency would, by necessity, appear in our conscious brains and how this fact could have contributed to the formation of complex societies. Next, investigating the Ego Tunnel during the dream state will give us even deeper insight into the conditions under which a true subject of experience emerges. How does the Dream Tunnel become an Ego Tunnel?

FIVE

✳

PHILOSOPHICAL PSYCHONAUTICS

What Can We Learn from Lucid Dreaming?

During the night of May 6, 1986, I became consciously aware that I was sleeping and also spiraling out of my physical body, in the typical manner described by Swiss biochemist Ernst Waelti (see chapter 3). Here is my "case study":

> Standing in front of my bed, I immediately realized that, for the first time in two years, I had entered the OBE state again. The clarity, the same electrified sense of lightness in my double body, made me excited and extremely happy, and I immediately began to experiment. I moved toward the closed glass door of the second-floor balcony in my parents' house. I touched the door, gently pushing it until I penetrated it and slid out onto the balcony. I flew down into the garden and landed on the lawn, where I moved around in the dim moonlight and looked at things. Again, the overall experience was crystal clear.
>
> When I became afraid of not being able to sustain the condition much longer, I flew back up, somehow returned to my physical body, and awoke with a mixture of great pride and joy. I had not managed to make any verifiable observations, but I had

had another OBE, in a clear, cognitively lucid way, fully controlled and without any intermediary blackouts. I sat up, wanting to take notes as long as everything was still fresh, but couldn't find a pencil.

I jumped out of bed and went over to my sister (who slept in the same room), woke her up, and told her, with great excitement, that I had *just* managed to do it again, that I had *just* been down in the garden, bouncing around on the lawn a minute ago. My sister looked at her alarm clock and said, "Man, it's quarter to three! Why did you have to wake me up? Can't this wait until breakfast? Turn out the light and leave me alone!" She turned over and went back to sleep. I was a bit disappointed at this lack of interest.

I also noticed that while fumbling with the alarm clock, she had accidentally set it off. It was beeping away and I hoped it hadn't wakened anybody else. Too late! I could hear someone approaching.

At that moment, I woke up. I was not upstairs in my parents' house in Frankfurt but in my basement room, in the house I shared with four friends about thirty-five kilometers away. It was not quarter to three at night; the sun was shining and I had obviously been taking a short afternoon nap. For more than five minutes, I sat on the edge of my bed almost frozen, not daring to move. I was unsure how real *this* situation was. I did not understand what had just happened to me. I didn't dare move, because I was afraid I might wake up again, into yet another ultrarealistic environment.

In dream research, this is a well-known phenomenon called *false awakening*. Did I really have an out-of-body experience? Or did I only have a lucid dream of an out-of-body experience? Can one slide from an OBE into an ordinary dream via a false awakening? Are all OBEs forms of lucid dreaming in the first place? To wake up twice in a row is something that can shatter many of the theoretical intuitions you have about consciousness—for instance, that the vividness, the coherence, and the crispness of a conscious experience are evidence that you are really in touch with reality. Apparently, what we call "waking up" is something that can happen to you at any point in phenomenological time. This is a highly relevant empirical fact for philosophical epistemology. Do you recall from chapter 2 the discussion about the evolution of human consciousness and how the distinction between things that only appear to us and objective fact became an element of our lived reality? Now we can see what it means that the appearance/reality distinction emerged only on the level of appearance: False awakenings demonstrate that consciousness is never more than the appearance of a world. There is no certainty involved, not even about the state, the general category of conscious experience in which you find yourself. So, how do you know that you actually woke up this morning? Couldn't it be that everything you have ever experienced was only a dream?[1]

Dreams are conscious because they create the appearance of a world, but, as noted in chapter 2, they are offline states—global states of conscious experience in which the Ego is decoupled from sensory input and unable to generate overt motor behavior. The dream tunnel not only contains the appearance of a world but also (in most cases) creates a fully embodied, spatially extended self moving around in a spatially extended environment. The virtual self thus born is an exclusively internal phenomenon in an even stronger sense than that of the waking self: It is immersed in a dense mesh of causal relations, all of which are internal to the brain. Dreamers are self-aware, but functionally they are not *situated*. Dreams are subjective states in that there is a phenomenal self; however, the perspective from which this conscious self perceives the world is very different—and much more unstable—than it is during wakefulness.

Have you ever noticed that you cannot control your attentional focus in your dreams? High-level attention is typically missing. Accordingly, the dream-self generated inside the Ego Tunnel when you are sleeping lacks the specific phenomenal quality I described in the preceding chapter as *attentional agency*, the conscious experience of directing the beam of your inner flashlight deliberately and selectively at various objects. But attentional agency is not just the ability to "zoom in" on certain things or point your mind at particular features of your world-model; it also entails the sense of ownership—ownership of the selection process preceding the shift in attention. Both aspects are missing in dreams. In a way, you are like an infant or a severely intoxicated person. The dream Ego is much weaker than the waking Ego.

If one penetrates deeper into the specific phenomenology created by the dreaming Ego, one discovers a considerable weakness of will and severe distortions of the thought process. In ordinary dreams, you sometimes cannot experience yourself as any sort of agent at all. It is difficult, for example, to make a decision and follow through with it. But even if you manage that, you are typically unable to ascribe agency to yourself. The dreaming self is a confused thinker, severely disoriented with regard to places, times, and people's identities. Short-term memory is greatly impaired and unreliable. Also, only rarely does the dream self have such sensory experiences as pain, temperature, smell, or taste. Even more interesting is the extreme instability of the first-person perspective: Attention, thinking, and willing are highly unstable and exist only intermittently, yet the ordinary dreaming Ego does not really care about this, or even notice it. The dream self is like the anosognostic patient, who lacks insight into a deficit following brain injury.

At the same time, the dream self creates intense emotional experiences—some aspects of the self are clearly stronger in the dream tunnel than in the tunnel of waking consciousness. Anyone who has ever had a nightmare knows how intense the feeling of panic can become during dreams. In the dream state, the emotional self-model can be characterized by unusually intense degrees of feeling, though this is not true for all emotions; for example, fear, elation, and anger predominate over sadness, shame, and guilt.[2]

Occasionally, the dream tunnel enables the Ego to access information about itself that is unavailable during the waking state. Whereas short-term memory is commonly impaired, long-term memory can be greatly enhanced. For instance, it is possible to relive childhood episodes vividly—memories that would never have been accessible during wakefulness. We tend to forget these afterward, because most of us have weak dream recall. But as long as the dream lasts, we have access to state-specific forms of self-knowledge.

Blind people are sometimes able to see in dreams. Helen Keller, who turned blind and deaf at the age of nineteen months, emphasized the importance of these occasional visual experiences: "Blot out dreams, and the blind lose one of their chief comforts; for in the visions of sleep they behold their belief in the seeing mind and their expectation of light beyond the blank, narrow night justified."[3] In one study, congenitally blind subjects produced dream drawings that judges were unable to distinguish from drawings of sighted subjects, and as EEG correlates between them were sufficiently similar, this strongly suggests that they can see in their dreams—but do they?[4] It is also interesting to note that Keller's dream tunnel contained the phenomenal qualities associated with smell and taste, which most of us experience only rarely in the dream state. It seems as if her dream tunnel became richer because her waking tunnel had lost some of its qualitative dimensions.

The dream tunnel shows to what extent conscious experience is a virtual reality. It internally simulates a behavioral space, a space of possibilities in which you can act. It simulates real-life sense impressions. As discussed in chapter 3, this is exactly what modern designers of virtual realities are trying to achieve (indeed, one of the best scientific journals on virtual-reality technology is titled *Presence*). It is precisely this sense of presence and full immersion that our biological ancestors achieved long ago. The resultant Ego, however, has created a more robust sense of presence for dreaming *and* for waking life as well. If it had not done so, we probably would not be trying to create virtual realities today, nor would we research the ability of the human brain to achieve this miracle within itself.

Even though dreams are behavioral spaces, they are not causally coupled to the real behavioral space of the dreaming human organism. Dreamers are not bodily agents; their behavior is internal, simulated behavior. The inhibition of the spinal motor neurons prevents bodily behavior from being generated during dream sleep—that is, REM (rapid-eye-movement) sleep. This is how the dream Ego is separated from the physical body. When this motor inhibition fails, as it does in a disorder known as RBD (for "REM-sleep behavior disorder"), internal dream behavior is acted out in the waking world. Typically found in men

over sixty, RBD is associated with a loss of the muscle atonia that typically accompanies REM sleep. Patients suffering from RBD are forced to
act out dramatic and often violent dreams. They will shout or grunt.
They may attempt to strangle their bed partners, set fire to their beds,
jump out of windows, even fire a gun.[5] Later they will recall little or
nothing of this physical activity—unless they fall out of bed or bump
into furniture or injure themselves or someone else and wake up. But
they can usually recall the dreams themselves, which typically involve
such physical activities as fighting, running, chasing or being chased,
and attacking or being attacked. These patients also seem to experience
violent and aggressive dream content more frequently than healthy
subjects do. Obviously, this is a dangerous condition that can lead to
self-inflicted injuries and serious sleep deprivation. What we can learn
from it is how the dream body, in normal circumstances, is decoupled
from the physical body. Normally, dreamers are not bodily agents, and
all their behavior is purely internal, simulated behavior. But when motor
inhibition fails, as it does in RBD, internal dream behavior is enacted by
the physical body.

The most interesting feature of ordinary dreams leads to some
deeper philosophical considerations about the nature of consciousness.
The dream tunnel is generated in a very special configuration: During
REM sleep, as noted, there is an output blockade, responsible for the
paralysis of the sleeper, and there is an input blockade, which prevents
(at least to a degree) sensory signals in the sleeper's environment from
penetrating conscious experience. At the same time, chaotic internal
signals are generated by what are known as PGO waves. They are electrical bursts of neural activity named for the brain areas involved (the
pons, the lateral geniculate nucleus in the thalamus, and the occipital
primary visual cortex) and are closely related not only to eye movements but also to the processing of visual information.[6]

As the brain tries to understand and interpret this chaotic internal
pattern of signals, it starts telling itself a fairy tale, with the dream ego
playing the leading role. The interesting point is that the dream Ego does
not know that it is dreaming. It does not realize the signals it is turning
into an internal narrative are self-generated stimuli—in philosophical

jargon, this feature of the dream state is a "metacognitive deficit." The dream Ego is delusional, lacking insight into the nature of the state it is itself generating.

LUCID DREAMING

The natural question to ask is if there are any dreams with additional insight, dreams in which the dream self-model has become so strong and rich that it allows us to understand what is happening. Can one consciously enjoy one's own internal virtual reality? Is it possible to dream without the metacognitive deficit? The answer is yes. You can have dreams in which you are not only aware of the fact that you are dreaming but also possess a complete memory both of your dream life and your waking life, as well as the phenomenal property of agency on the levels of attention, thought, and behavior. Such dreams are called *lucid dreams*. They are highly interesting—not so much for the sheer fun of the drama but because they open new ways of investigating the phenomenon of conscious experience. In particular, they help us understand how the various layers of the self-model are constructed and woven into the dream tunnel.

Dutch psychiatrist Frederik van Eeden, who coined the phrase "lucid dreaming," reported the following experience to the Society for Psychical Research in 1913:

> In January, 1898 . . . I was able to repeat the observation. . . . I
> dreamt that I was lying in the garden before the windows of my
> study, and saw the eyes of my dog through the glass pane. I was
> lying on my chest and observing the dog very keenly. At the
> same time, however, I knew with perfect certainty that I was
> dreaming and lying on my back in my bed. And then I resolved
> to wake up slowly and carefully and observe how my sensation
> of lying on my chest would change to the sensation of lying on
> my back. And so I did, slowly and deliberately, and the transi-
> tion—which I have since undergone many times—is most won-
> derful. It is like the feeling of slipping from one body into

another, and there is distinctly a *double* recollection of the two bodies. I remembered what I felt in my dream, lying on my chest; but returning into the day-life, I remembered also that my physical body had been quietly lying on its back all the while. This observation of a double memory I have had many times since. It is so indubitable that it leads almost unavoidably to the conception of a dream-body.[7]

Van Eeden's "dream-body" is the self-model in the dream state. Lucid dreams are fascinating because our naive realism—our unawareness of living our lives in an Ego Tunnel—is temporarily suspended. They are therefore a promising route of research for solving what I termed the Reality Problem in our tour of the tunnel in chapter 2. A lucid dream is a global simulation of a world in which we suddenly become aware that it is indeed just a simulation. It is a tunnel whose inhabitant begins to realize that he or she actually operates in a tunnel all the time.

Hugh G. Callaway, a British experimenter in out-of-body experiences who published under the pseudonym Oliver Fox, recorded the following classic episode, occurring in 1902, when he was a young science student at the Harley Institute at Southampton:

I dreamed that I was standing on the pavement outside my home. . . . I was about to enter the house when, on glancing casually at [the pavement] stones, my attention became riveted by a passing strange phenomenon, so extraordinary that I could not believe my eyes—they had seemingly all changed their position in the night, and the long sides were parallel to the curb! Then the solution flashed upon me: though this glorious summer morning seemed as real as real could be, I was *dreaming*! With the realization of this fact, the quality of the dream changed in a manner very difficult to convey to one who has not had the experience. Instantly, the vividness of life increased a hundredfold. Never had the sea and sky and trees shone with such glamorous beauty; even the commonplace houses seemed alive and mystically beautiful. Never had I felt so absolutely well,

so clear-brained, so inexpréssibly *free*! The sensation was exqui-
site beyond words; but it lasted only a few minutes and I awoke.[8]

Maybe you've had a lucid dream yourself; the phenomenon is not rare. If
not, you can try a number of different induction techniques. For instance, you can adopt the habit of performing "reality checks" several
times a day. Each reality check should last at least a minute. It consists in
carefully inspecting your current inner and outer environment for any
indications that this might not be ordinary waking reality. Here is a
checklist that readers interested in exploring the dream tunnel can use
as a guideline.

- Is all the furniture arranged the way it normally is?
- Are the paving stones, the tiles, or the design of the carpet on the
  floor arranged in the same pattern as usual?
- Do objects or persons suddenly appear and disappear, or do they
  change their identity?
- Do you know who you are and where you are?
- Do you remember what day of the week it is and when you last
  woke up?
- Are there any gaps in your short-term memory of recent events?
- Does your visual attention shift the way it usually does?
- Are you engaging in any unusual physical activities, like flying?
- Are you constantly trying to remember something you know is of
  great importance but can't remember what it is?
- Does your current situation have a metaphoric or symbolic
  character, or do you have the feeling of being close to an
  important discovery?

If you perform reality checks of this type several times a day, you have a
good chance of eventually becoming a lucid dreamer. By pure habit, you
will one day perform a reality check in a dream—and if you are lucky,
you will correctly realize that you are dreaming.[9]

Other methods of inducing lucid dreams are even more efficient. Try
setting an alarm clock early in the morning and carefully writing down

the events of your last dream. Get up, move around for a short period of time, and then go back to bed. While you are falling asleep, try to rehearse the last sequence of dream events in as much detail as you can. You may find that you can consciously reenter the dream and stay lucid throughout it.[10]

As an intrepid philosophical psychonaut, I have of course tried to build devices to do this kind of exploring, involving headphones and tape loops whispering, "Watch out—this is a dream!" at thirty-minute intervals all night long. I also bought an expensive lucid-dreaming device called a Nova Dreamer, which looks a bit like the eye masks you sometimes see people wearing on long-distance plane flights. The Nova Dreamer is activated when your rapid eye movements signal the start of a dream. After a couple of minutes, it begins submitting mild subliminal visual stimuli, and you can perceive these soft, red, ring-shaped flashes of light through your closed eyelids. They are meant to alert you to the fact that you are dreaming; however, they are more likely to be integrated into your dream story. Here's one of my own dreams thus invaded:

> I am an astronaut. I have been waiting for this moment for years. My friend and I are lying on our backs in the Space Shuttle, awaiting takeoff with a mixture of anxiety and great excitement. Deep below our backs, we can feel the rumbling and rattling of the ignition give way to a thundering roar. Then red lights start flashing everywhere on the control panel. Suddenly, every possible alarm system is activated. Someone says, "Something must have gone terribly wrong!" We feel the spaceship slowly tilting to the side and losing its vertical position while the roar at our backs gets louder and louder.

Unfortunately, all I ever got out of my expensive lucid-dreaming device was terrible nightmares—with an interesting twist. In Germany, the flashing lights of police cars are blue. So what I got from this device was American nightmares, with American police cars hunting me down and cornering me, flashing red lights and all. Every two years or so, I give my Nova Dreamer another try; lately, it has had a different effect on me. I wake up in the morning and the device is gone. If I go looking for it, I find that it has been hurled across the bedroom by some stranger. Apparently there is someone inside me who does not want to be a philosophical psychonaut or a serious practitioner of first-person phenomenological research at all—someone who just wants to sleep.

So what, exactly, is a lucid dream? In a lucid dream, the dreamer knows that she is currently experiencing a dream and is able to ascribe this property to herself. If we opt for a strong definition, another condition is that she also has access to memories of her previous dream and waking life. Autobiographical memory is fully intact. The dreamer has full access not only to past conscious experiences in waking life and in ordinary dreams but also to previously experienced lucid dreams. The overall level of mental clarity and cognitive insight is at least as high as it is during normal waking states. A further defining characteristic is that, according to subjective experience, all five senses function just as well as they do during the waking state. Finally, and perhaps most important, the property of *agency* is fully realized in the lucid dream. Phenomenologically, the lucid dreamer knows about her freedom of will. Not only can she direct the focus of attention wherever she likes, but she can also actually *do* whatever she wants—fly, walk through walls, or engage in conversations with dream figures. The subject of a lucid dream is not a passive victim lost in a sequence of bizarre episodes but rather is a full-blown agent, capable of selecting from a variety of possible actions.

Full control of one's attention is an important feature distinguishing lucid dreams from ordinary dreams. Insight into one's freedom to act is also an important criterion of lucid dreams (but is it an insight?). During what are sometimes called *pre-lucid dreams*, we frequently become aware that none of this is real, that this must be a dream, but we remain passive observers. With the onset of full lucidity, the dreamer often turns from a passive observer into an agent—someone who takes charge, moves around, explores and experiments, who deliberately starts to interact with the dream world and shape it.

My favorite experiment in lucid-dream research was conducted by Stanford University psychophysiologist Stephen LaBerge and his colleagues more than a quarter of a century ago.[11] It exploited the interesting fact that our conscious self-model is firmly anchored in the brain in a fascinating way: There is a direct and reliable relationship between the gaze shifts reported by lucid dreamers and the eye movements observed in their sleeping bodies. In the sleep lab, these eye movements can be recorded using a polygraph. The fact that the movements of the dream-eyeballs in the dream-body are directly correlated with the movements of the physical eyeballs in one's physical body was used by LaBerge in a particularly ingenious experiment. Veteran subjects deliberately indicated the initiation of a lucid dream with specific ocular signals determined before the experiments—that is, by rapidly moving their eyes up and down. Two such eye shifts would inform the experimenters of the onset of a lucid dream; four signaled awakening. The polygraphic analysis revealed that the onset of lucidity is typically correlated with the first two minutes of an REM phase, or with short intervals of waking consciousness during an REM phase, or with heightened phasic REM activity (characterized by bursts of eye movements and sometimes by motor twitches and widespread synchronized activity in specific thalamocortical networks).[12] Put simply, lucidity seems to occur when there is a brief and sudden increase of the general cortical level of arousal: All nerve cells become more active, the result being the sudden availability of more "computational power," or capacity for information processing. With regard to the dream itself, lucidity seems to lead to increased vividness, heightened fear or stress, the discovery of contradictions within the dream world, and, of course, the subjective experience of becoming aware of a "dreamlike" or "unreal" quality of reality.

I like these experiments because they are a rare example of trans-tunnel communication. When the lucid dreamer in the sleep lab emits eye signals by deliberately moving his or her dream-eyes up and down and scientists in the waking world read these signals off their instruments, a multiuser link between the dream tunnel and the waking tunnel is established. Because the gaze shifts performed by the dream-body are functionally linked to those of the physical body, and because the lucid dreamer is aware of this fact, a bridge connecting the two tunnels is built. In this experimental setup, information from one type of conscious reality tunnel can be transmitted to another type, one created by the brains of other human beings.

We need more good empirical research on lucid dreams. It is plausible to assume that lucidity depends on the degree to which the prefrontal cortex, where the organizing of cognitive and social behaviors takes place and the so-called executive functions are located, can form a stable functional link with other brain regions that generate the conscious dream self. The prefrontal cortex is thought to arrange thoughts and actions in accordance with internal goals. It also has to do with differentiation among conflicting thoughts, planning, assessing future consequences of current activities, predicting outcomes, generating expectations, and the like.

Allan Hobson, a psychiatrist and dream researcher at the Massachusetts Mental Health Center and author of *The Dreaming Brain,* has speculated that for lucidity to occur, "the normally deactivated dorsolateral prefrontal cortex (DLPFC) must be reactivated, but not so strongly as to suppress the pontolimbic signals to it."[13] This part of the brain may allow us to refer to ourselves by engaging in reflective thought. In the lucid-dream tunnel, this leads to the reestablishment of executive control and the reemergence of a full-blown agent. If Hobson is right, the moment we consciously think, "My God, I'm dreaming!" may be the moment the self-model of the dream state becomes hooked up to the prefrontal cortex, making proper reflexive self-consciousness possible again and reestablishing cognitive agency.

Here are some questions for future research: What precisely happens to the conscious self during the transition from an ordinary dream to a lucid dream? What are the fine-grained functional differences between the dream self-model and the lucid self-model? Could there also be something like "lucid waking"? And what, exactly, happens during a false awakening?

As we have seen, false awakenings can happen to all of us. This brings up another classical philosophical problem, the issue of solipsism (from Latin: *solus,* alone, and *ipse,* self). How, exactly, can I refute the skeptical hypothesis that my mind is the only thing that I know to exist?

How can I exclude the possibility that the external world—and other conscious minds in particular—cannot be known and might not exist at all? Finally, here is a little thought experiment in applied tunnel epistemology, introduced and illustrated by a lucid dream reported by the late German dream researcher Paul Tholey:

> I briefly looked back. The person following me did not look like an ordinary human being; he was as tall as a giant and reminded me of *Rübezahl* [a mountain spirit in German legend]. Now it was fully clear to me that I was undergoing a dream, and with a great sense of relief I continued running away. Then it suddenly occurred to me that I did not have to escape but was capable of doing something else. I remembered my plan of talking to other persons during the dream. So I stopped running, turned around, and allowed the pursuer to approach me. Then I asked him what it actually was that he wanted. His answer was: "How am I supposed to know?! After all, this is *your* dream and, moreover, *you* studied psychology and not me."[14]

Imagine that while in the dream tunnel, you suddenly become lucid and find yourself at a major interdisciplinary conference, where dream scientists and dream philosophers are discussing the nature of consciousness:

> While they're standing around during the coffee break, one of them claims that *you* do not really exist, because you're just a dream figure in your own lucid-dream tunnel, a mere possibility. Amused, you reply, "No, *you* are all figures in my dream— just figments of my imagination." This response is greeted with laughter, and you notice, too, that colleagues at other tables are grinning and turning their heads in your direction. "All of this is happening in *my* brain!" you insist. "I own the hardware, and you are all just simulated dream characters in a simulated environment, processed and created by my central nervous system. It would be easy for me. . ." Here, more laughter interrupts you—roars of laughter. A young PhD student arrogantly starts

explaining the basic assumptions about the nature of reality shared by this particular scientific community: No such things as brains or physical objects ever existed. The *contents* of consciousness are all there is. So all phenomenal selves are equal. There is no such thing as an individual "tunnel" in which one self-model represents the true subject of experience and all other person-models are just dream figures.

The strange philosophical concept this dream community of scientists has developed as their background assumption is known as *eliminative phenomenalism*. As the slightly overambitious PhD student explains: "Eliminative phenomenalism is the thesis that physics and the neuroscientific image of man constitute a radically false theory, a theory so fundamentally defective that both the principles and the ontology of that theory will eventually be displaced, rather than smoothly reduced, by a completed science of pure consciousness." All reality, accordingly, is phenomenal reality. The only way you can drop out of this reality is by making the grandiose (but fundamentally false) assumption that there actually is an outside world and that you are the subject—that is, the experiencer—of this phenomenal reality, that there actually is a consciousness tunnel (a *wormhole*, as they ironically call it), and that it is your *own* tunnel. By entertaining this belief, however, you would suddenly become unreal and turn into something even less than a mere dream figure yourself: a *possible* person—exactly what your opponent claimed at the beginning of the discussion.

"Listen, guys," you say, in a slightly irritated voice, "I can demonstrate to you that this is *my* consciousness tunnel, because I can end this state, and your very existence, at any point in time. A well-known technique for terminating lucid dreams is to hold one's hands up in front of one's eyes and fix one's visual attention on them. If I do this, it will interrupt the rapid eye movements in my physical body and thus end the dream state in my physical brain. *I* will wake up in the Waking Tunnel. *You* will simply cease to exist. Do you want me to show you?" You note that your tone of voice sounds triumphant, but you also note that the amusement in the eyes of the other scientists and philosophers has changed to pity. The arrogant PhD student blurts out again: "But don't you see that simply falling back into what you call 'waking up' doesn't prove anything to anybody? You must demonstrate the truth of your ontological assumptions to *this* scientific community, on *this* level of reality. You cannot decide the question by simply degrading yourself to a virtual person and disappearing from *our* level. By waking up, you will learn nothing new. And you cannot prove anything at all—certainly not to us, but not to yourself, either. If you want to humiliate yourself by vanishing into your waking wormhole, then just go ahead! But the serious pursuit of consciousness research and of philosophical theory of science is something entirely different!"

How would you react? If I had not made the right decision at this point, I might never have finished this book. But enough tunnel epistemology for now.

CHAPTER FIVE APPENDIX
DREAMING: A CONVERSATION WITH ALLAN HOBSON

Allan Hobson is professor of psychiatry at Harvard Medical School, where he founded the Laboratory of Neurophysiology in order to study the brain basis of dreaming.

Working with Dr. Robert McCarley, Hobson developed the reciprocal interaction model, according to which REM (rapid eye movement) sleep is generated by cholinergic brainstem mechanisms, and the activation-synthesis theory, which views dreaming as the result of automatic brain activation and the synthesis of chaotic internal signals during sleep. In the course of his extensive experiments with human sleep laboratory data, Hobson invented the Nightcap method of home-based sleep recording and, with Robert Stickgold, used this method to characterize conscious states around the clock. Hobson and Stickgold also developed a new approach to the study of sleep effects on learning.

Recently, Hobson has integrated his own ideas and findings with new data coming from PET and lesion studies of human sleep in a general model of state-dependent aspects of consciousness. Called AIM, the new model maps three dimensions—activation (A), input-output gating (I), and chemical modulation (M)—to define a state space through which the brain-mind travels in a recurrent loop as we wake, sleep, and dream.

Hobson is the author of many books, including *The Dreaming Brain* (1989), *Sleep* (1995), *Consciousness* (1999), *Dreaming as Delirium: How the Brain Goes Out of Its Mind* (1999), *The Dream Drugstore* (2001), *Out of Its Mind: Psychiatry in Crisis* (2001), and *13 Dreams Freud Never Had: The New Mind Science* (2004).

Metzinger: What exactly is special about consciousness in the dream state, compared to consciousness in wakefulness and non-REM sleep?

Hobson: Dream consciousness is more intense, more single-minded, more elaborate, and more bizarre than consciousness in waking. Hence, it can reasonably be viewed as the most autocreative state of the brain-mind. It is also the most psychosis-like state of normal consciousness. Because its neurobiology is so well known, its study offers us a unique scientific opportunity to understand ourselves better in both health and disease.

Metzinger: And what exactly is the relationship between REM sleep and dreaming?

Hobson: The correlation is quantitative, not qualitative. Dreamlike mental activity is also correlated with sleep onset (stage I) and with late-night sleep (stage II), but at all times of the night or day, the correlation is highest in REM. As for the actual relationship, my hypothesis is that dreaming is our subjective awareness of our brain activation in any state of sleep. Activation is highest in REM sleep. So is dreaming. I think that dreaming and REM sleep are our subjective and objective references to the same fundamental process of the brain-mind. I am a monist, through and through. How about you?

Metzinger: Sure—I have always liked philosophers like Spinoza, Bertrand Russell, or Herbert Feigl, who were neutral monists and thought that the distinction between physical and psychological states is actually quite superficial and rather uninteresting. For us philosophers, the more important problem, of course, is what *precisely* "through and through" means. But right now, you are the one that has to answer difficult questions! So, how do you explain the relationship between chaotic dream content, generated by the brainstem, and the more nonrandom and seemingly meaningful aspects of dreaming?

Hobson: Be careful, Thomas, or you will fall into the "either/or" trap that has swallowed up so many of our distinguished colleagues. The answer is "both/and." REM sleep is generated by the brainstem, while dreaming is the subjective experience of the brainstem's activation of the forebrain in REM sleep. The REM generation process has many chaotic features, which the forebrain tries its best to integrate into a coherent story. But the forebrain is also in a different state than it is in waking, which makes its job more difficult. The forebrain does the best it can under difficult circumstances. Whether you think it does well or not so well depends on whether you think the cup is half empty or half full. Both are true.

Metzinger: Which parts of the human brain are absolutely necessary for dreaming? Without which parts is it impossible to dream?

Hobson: The second question has empirical evidence contributing to an answer, but the first question is much more interesting and much more complex. Unfortunately, it cannot be answered scientifically.

Take the second question first. The neuropsychologist Mark Solms asked some three hundred stroke patients whether they had noticed any change in their dreaming after their strokes. Patients reported a complete cessation of dreaming if their stroke damaged either the parietal operculum or the deep frontal white matter. These claims were particularly interesting, because these same brain regions were selectively activated in PET studies of REM sleep. Another finding of interest is the report of dream cessation after prefrontal lobotomy, which Solms discovered in the literature of the 1940s and 1950s.

On their face, these findings suggest that dreaming depends upon the brain's capacity to integrate emotional and sensory data when activated offline. But of course this doesn't answer the first question at all. Many other brain regions are likely to be equally essential to dreaming. For example, the visual system must be involved—and, indeed, Solms's patients reported the loss of visual imagery in their dreams if their strokes affected the occipital cortex. Presumably, the loss of dreaming is an example of what Norman Geschwind called disconnection syndrome. In other words, the damaged areas are cerebral crossroads which, when damaged, prevent other parts of the brain from interacting properly. The important role of the brain-stem is unlikely to be revealed by this technique, since lesions large enough to impede dreaming are likely to be fatal or lead to unre-sponsive vegetative states.

There are several problems with this approach to dream science. The first is that the answer to question two does not answer ques-tion one. It is possible, for example, to imagine that Broca's and Wer-nicke's areas may be quite important to the confabulatory quality of dreaming, but this possibility cannot be tested if the patient has lost his ability to give dream reports! Furthermore, it is important to point out that all of Solms's data deal with dream reporting, which cannot be equated with dreaming. In fact, most of us have little or no memory of our dreams.

In Solms's studies and in the earlier works of [Cristiano] Violani and [F.] Dorrichi and of [M. J.] Farah and [M. S.] Greenberg, which reached similar conclusions about the parietal operculum, there was no effort made to record the patients' sleep or to wake them up to elicit dream reports. These controls are important and yet to be per-formed. Solms and others are to be congratulated for opening up the neuropsychological study of dreaming. We look forward to learning more from this approach. For the present, all we can say is that dreaming depends on the selective activation and deactivation of many brain regions, including those which, when damaged, lead to the failure to report dreams.

Metzinger: What do you think was most likely the evolutionary func-tion of dreaming, and when did it first develop?

Hobson: Regarding evolution and the functional advantages of having a brain that can dream, I have both conservative and speculative views. The conservative position is that there is no evidence that dreaming itself serves any purpose whatsoever. That is to say, neither the conscious awareness of dreams while they are occurring nor recall of such awareness on awakening from sleep is likely or has been shown to be useful. I think we must take seriously Owen Flanagan's suggestion that dreams are the spandrels of sleep.[15]

At its most extreme, the argument says that dream consciousness is an epiphenomenon, which humans and other animals can do just as well without. The most cogent reason for thinking this may be true is the all-but-complete amnesia that we have for our dreams. If dream recall were adaptive, surely we would have more of it! But taking this position about dreaming as a conscious experience does not negate a healthy, speculative interest in the functional significance of having a brain that can self-activate in sleep. Such a brain could be doing many things. These include the already known enhancement of motor learning, the regulation of dietary and thermal calories, and the improvement in the immune functions. I don't have to be aware of those functions, even if they are essential to my survival and my reproductive success.

Here we are at the nub of a number of critical philosophical questions, including the common confusion of brain activity and awareness. Our conscious awareness during waking is an obvious adaptive advantage, but our conscious awareness during sleep may not be. It may even be an adaptive advantage *not* to remember dream content. Allowing for some therapists' assertion that dreaming is the royal road to the unconscious, it is still possible to ask, "Who wants to go there?" Those who do are free to try, but I myself see no adaptive advantage to dream recall and dream interpretation, even though I myself indulge, with great pleasure, in both sports.

My own special theory is that dreaming is a highly distinctive form of conscious awareness that can be used to better understand the brain activity that leads to consciousness, whether it be in waking or in sleep. As Gerald Edelman and Giulio Tononi have pointed out, it is the vast thalamocortical system that must be activated to produce consciousness. In waking and in sleep, this system is activated by the brainstem, but the chemical modulation accompanying the activation is very different in the two states. The contributions of other structures, like the limbic system and the modulatory systems of the brainstem, are very significant in that they "color" consciousness as well as activate it.

Humans and most other mammals have brains that can self-activate in sleep, when environmental conditions such as cold and darkness do not favor waking behavior, and it is this capacity, not the awareness of it, that is significant for evolutionary success.

Metzinger: What is known today about the phylogenetic manifestation of the sleep-wake cycle? How did it come about in our ancestors? And how is that manifestation related to consciousness?

Hobson: The answer is that a lot is known! Without going into elaborate details, it can be safely said that the fully developed sleep-wake cycle, with alternative phases of NREM and REM sleep, is an adaptation reserved to homeothermic animals—namely, mammals and birds that regulate their body temperature. What is the adaptive link between homeothermia and sleep? Again, the answer is simple. Keeping brain temperature constant despite enormous fluctuations of environmental temperature guarantees reliable brain function in a wide variety of environmental contexts. In other words, temperature control and brain function are tightly linked, and sleep secures that link.

With respect to the consciousness angle, I follow Edelman, who refers to primary consciousness—that is, perception, emotion, and memory—and secondary consciousness, which is awareness of awareness and the ability to describe it. Secondary consciousness, which depends upon language and other sophisticated abstractions, is exclusively human. Primary consciousness is widespread among mammals and could even be present in some submammalian species. Unfortunately, these assertions can never be more than intelligent guesses, because no subhuman animal can communicate its subjective experience verbally. Animal-rights activists, like right-to-life agitators, are quite right in claiming that many subhuman and immature animals are, to a limited but significant degree, conscious. If we are to take their lives or cause them pain, we had better have a strong moral reason for doing so. And we do. It is the reduction of human suffering. I am an unapologetic human supremacist. Just as I take animal and vegetable life to survive, I take it to promote the quality of that life.

Metzinger: Could we build a machine that dreams but never wakes? Are there animals that dream but do not enjoy waking consciousness?

Hobson: Again, it's easier to answer the second part of the question. Given the limitations to scientific knowledge I've emphasized, the answer is no. If an animal can activate its brain in sleep, it has that capacity in waking also. So it stands to reason—but it is only reason—that animals that have (necessarily limited) dream consciousness also enjoy consciousness in waking. As for the first part, a dream machine can already be designed, but there is a state-of-the-art limitation that cripples the program. That limitation is the problem of generating linguistic statements from a biographical database. The last time I consulted with language specialist Roger Shanks, he told me that this crucial piece was still missing from the AI (artificial intelligence) puzzle. Activating perception and emotion modules poses no problem, and making them responsive to or independent of input and output can be done, as John Antrobus of CUNY has already shown. Any dream machine that would now be designed would likely have a wake-state mode of operation, because we're interested in the similarities and differences between those two states and how they are generated. But it's theoretically possible to develop a dream-only machine.

The fact that—as far as we know—evolution has not yet produced any dream-only animals suggests a deep meaningful and functional link between the waking and dreaming states of consciousness and brain activity. It is possible to argue, as I've already pointed out, that the brain is activated offline to benefit the brain online and vice versa, without postulating a causal link between the conscious awareness of the two states.

Metzinger: Cultural as opposed to biological evolution certainly gives a
    place to dream content, but whether that place is truly adaptive is
    still questionable.

Hobson: Many cultures have accorded prophetic meaning to dreams.
    The widely shared view of all such prophets is that dreaming is a
    message, in code, from important external or internal agents and
    needs decoding. Such decoding is seen by the practicing cultures as
    not only valid but also determinant of important conscious personal
    and political decisions. The dream sorcerers helped kings decide
    whether or not to go to war. Should modern psychoanalysts help in-
    dividuals decide, say, whether to pursue a relationship further or
    not, based on the patients' dreams?

    One problem with this approach is the religious belief that there
    is some hidden truth that only dreams can reveal. Thus, one mys-
    tery, dreaming, is used to explain another, decision making. There is
    no evidence that this belief is justified. As Adolf Grünbaum has
    shown in his discussion of the Tally Argument, customer satisfac-
    tion cannot be used as a scientific warrant for the truth-value of a
    prophetic assertion—or a dream-interpretation scheme.

    It might well be that dreaming reveals one's cognitive repertoires
    in dealing with emotion, but that is not particularly difficult to dis-
    cern in waking. The stronger claim, by psychoanalysis, that dream
    interpretation reveals *hidden* links between cognition and emotion,
    has no scientific proof whatsoever.

Metzinger: I am particularly interested in the transition between ordi-
    nary dreams and lucid dreams. What are the necessary and suffi-
    cient conditions in the brain for lucidity to occur? What exactly is
    the role of the dorsolateral prefrontal cortex?

Hobson: The occasional awareness that one is in fact dreaming is an ex-
    tremely informative detail of modern dream science. The fact that
    such insight can be cultivated thickens the plot considerably. Taken
    together, the data suggest that the conscious state accompanying
    brain activation in sleep is both plastic and causal. It is plastic be-
    cause self-reflective awareness occasionally does arise sponta-

neously, and because with practice its incidence—and its power—can be increased. It is causal because lucidity can be amplified to command scene changes in dreams and even to command awakening, the better to remember, and enjoy, occasional dream-plot control. My position about lucidity is that it *is* real, it *is* powerful, and it *is* informative.

With respect to the third point, we already know, thanks to Stephen LaBerge, that sleep lucidity occurs in REM sleep, and we can predict that during lucid REM sleep dreaming, the dorsolateral prefrontal cortex, or DLPFC, which is selectively deactivated, may become reactivated so that the ponto-thalamical show of dreams comes under conscious control. I believe that this hypothesis, which is testable, contains the answer to many fundamental neurobiological and philosophical questions, such as the relationship of brain activity to consciousness and the causality of consciousness—free will.

If, as I predict, the DLPFC does reactivate during lucid dreaming while the ponto-thalamocortical dream show continues, then Daniel Dennett's despised Cartesian theater *does* exist. One part of the brain—the seat of the executive ego—wakes up and watches, or even directs, the dream show thrown up on the consciousness screen by the activation of the pons, thalamus, cortex, and limbic system. Eat your heart out, Daniel Dennett!

The evanescence and fragility of the lucid dream state testify to its unlikelihood and its nonadaptive nature. The lucid dream also demands the special attention that all such revelatory rarities deserve. Unfortunately, it is unlikely to get that attention. The reason is that the experiments will be difficult to perform and expensive to underwrite. This would be a barrier to many more trivial exercises in cognitive neuroscience, but lucid dreaming has a bad name because (a) many scientists still do not believe it is real, (b) many do not trust LaBerge's data about its occurrence in REM sleep, and (c) many will not go near the lucid-dream problem, because they fear being labeled as cranks or nuts! You, Thomas Metzinger, should easily be able to understand this fear.

Metzinger: Well, I certainly know what you're talking about. The right strategy would be not to declare such areas taboo but to invade them with open-minded, unbiased scientific rationality. The problem in the background, of course, is that if we want to be realistic, we also have to admit that the newly emerging field of consciousness research is not populated by philosophical saints interested in the pursuit of self-knowledge as such. It is driven at least as strongly by what I sometimes call the Teflon-coated Darwin machines of Academia—brute individual career interests. Scientists, of course, are self-sustaining, risk-avoiding Ego Machines as well. Sad to say, the field of lucid-dream research is not moving well at this time.

Hobson: In my opinion—which is not widely shared, even by Thomas Metzinger—we need to work on a science of subjectivity. In order to be able to utilize first-person data, we need to be both cautious and versatile. Reports of conscious experience must be collected from many individuals in many states. These reports must be rigorously quantified, and the states with which they are associated must be objectified. The brain states must be more fully characterized using a full panoply of techniques, including PET and MRI in humans, cellular and molecular probes in animals, behavioral tests in humans, and more.

Who will do all these things? As far as I know, I'm the only person in the world even to have tried. I say this with all due modesty and even sincere self-deprecation. I am proud of my accomplishments, but I can easily understand the criticism of my work as a fool's errand. At the root, the approach I advocate is in the service of the emergentist hypothesis of great scientists like Roger Sperry and great philosophers like William James. Such thinkers are few and far between.

More common, and far more handsomely rewarded, are those who dig deep discovering molecular widgets within and between nerve cells. Such discoveries are truly wondrous, but they will never lead to an understanding of conscious experience. Interestingly, even such well-known colleagues as Sigmund Freud worked within

this ill-fated reductionist paradigm. Here I use the term "reductionist" in its popular sense, implying eliminative materialism.

Metzinger: Why are you interested in philosophy? What contributions
from the humanities are you looking for?

Hobson: I'm interested in philosophy because I believe it is a founda-
tional discipline—along with psychology and physiology—of cognitive neuroscience as it tries to figure out how to study consciousness.
I myself try to "do" philosophy, but I need help. That's why I turn to
people like you, Owen Flanagan, and even David Chalmers. In general, I get positive responses from philosophers. They're genuinely
interested in my efforts and they generously share their insights with
me. You're no exception.

Regarding the second part of your question, I want philosophers
and other humanists to realize that the scientific study of brain-
mind states constitutes one of the greatest challenges and opportunities to better understand ourselves that has ever been presented
to us in our long intellectual history. There is room for many disciplines in this effort, which is as simple and broad-based as it is am-
bitious. To bring more coworkers up to speed is my own private
goal. We need all the help we can get. I even believe that brain-
mind science is one of the humanities.

Metzinger: So, today, is there still any point in psychoanalysis, or is it
just a lot of hot air? What do you think of Solms's arguments?

Hobson: Sigmund Freud was fifty-percent right and a hundred-percent
wrong! So is Mark Solms, but for different reasons. Freud was right
to be interested in dreams and what dreaming could tell us about
the human psyche, and especially its emotional aspects. His dream
theory is now obsolete, but its errors are still being promoted by
such psychoanalysts as Mark Solms.

Here's a checklist of Freudian hypotheses and the corresponding
alternatives offered by modern neurobiology:

(1) *Instigation of Dreaming.*
Freud: release of unconscious wishes.
Neurobiology: brain activation in sleep.

(2) *Characteristics of Dreaming.*
  (a) *Bizarreness.*
      Freud: disguise and censorship of unconscious wishes.
      Neurobiology: chaotic, bottom-up activation processes.
  (b) *Strong emotion.*
      Freud: Can't explain that one!
      Neurobiology: selective activation of limbic lobe.
  (c) *Amnesia.*
      Freud: repression.
      Neurobiology: aminergic demodulation.
  (d) *Hallucinations.*
      Freud: regression to the sensory side.
      Neurobiology: activation of REMs and PGO waves.
  (e) *Delusion, loss of self-reflective awareness.*
      Freud: ego dissolution.
      Neurobiology: selective deactivation of the dorsolateral prefrontal cortex.

(3) *Function of Dreaming.*
    Freud: guardian of sleep.
    Neurobiology: epiphenomenon, but REM sleep essential to life via enhancement of thermoregulatory and immune functions.

As we say in America, "Ya pays your money and ya takes your choice!" I choose neurobiology. How about you? As for Solms, he is nothing but a very smart psychoanalyst who wants to save Freud from the ashbin. His arguments, based on his important neuropsychological work, are weak. He has given up on disguise/censorship, but wants to resuscitate wish fulfillment. While it is true that dreams often do represent our desires, they are rarely truly unconscious, and dreams also represent our fears, a fact Freud could never explain. So what is left after Solms has given up disguise/censorship and only weakly defined wish fulfillment? Not much!

Solms attacks my activation-synthesis hypothesis of dreaming because of the observed dissociation between REM sleep and dreaming. As I have repeatedly pointed out, the correlation between REM sleep and dreaming is quantitative, not qualitative. The brain begins to shift from waking to REM sleep as soon as sleep begins. This means that the probability of dreaming begins to rise at sleep onset, to persist even in deep NREM sleep, when brain activation is still at eighty percent of waking levels, and to increase to its peak in REM sleep.

Why, then, do I say that Freud and Solms are fifty-percent right? Because dreams are not entirely nonsensical. They do make salient interrogations between emotions and cognition. Hence, they are worth reporting, discussing, and even interpreting, in terms of what they tell us about our emotions and how they influence our thoughts and our behavior. But they do this directly and openly, not via the symbolic transformation of forbidden wishes from the unconscious.

The good news is that you don't have to pay—or even leave your house—if you want to use dreams to explore your emotional life. You need only to pay attention, keep a journal, and reflect on the messages from your emotional brain, the limbic lobe. If you're a scientist, like me, you can do much more. You can use dreams and dreaming to build a new theory of consciousness.

SIX

.+.

THE EMPATHIC EGO

Have you ever watched a child who has just learned to walk run toward a desired object much too quickly and then trip and fall on his face? The child lifts his head, turns, and searches for his mother. He does so with a completely empty facial expression, showing no kind of emotional response. He looks into his mother's face to find out what has happened. How bad was it, really? Should I cry or should I laugh?

Toddlers do not yet have an autonomous self-model (though probably none of us has a self-model that is truly independent from others). In such small children, we observe an important fact about the nature of our own phenomenal Ego: It has social correlates as well as neural correlates. The toddler does not yet know how he should feel; therefore he looks at his mother's face in order to define the emotional content of his own conscious self-experience. His self-model does not yet have a stable emotional layer to which he could attend and, as it were, register the severity of what just happened. The fascinating point is that here are two biological organisms that just a few months ago, before being physically separated at birth, were one. Their Egos, their phenomenal self-models, are still intimately coupled on the functional level. When the toddler gazes at his mother and starts to smile in relief, there is a sudden transition in his PSM. Suddenly, he discovers that he didn't hurt himself at all, that the only thing that happened to him was a big surprise. An ambiguity is resolved: Now he knows how he *feels*.

There are kinds of self-experience that an isolated being could never have. Many layers of our self-model require social correlates; more than that, they are frequently created by some sort of social interaction. It is plausible to assume that if a child does not learn to activate the corresponding parts of his emotional Ego during a certain crucial period of his psychological development, he will not be able to have those feelings as an adult. We can enter certain regions in our phenomenal-state space only with the help of other human beings. In a more general sense, certain types of subjective experiences—interpersonal connectedness, trust, friendship, self-confidence—may be more or less available to each of us. The degree to which individuals have access to their emotional states varies. The same is true of their capacity for empathy and the ease with which they can read the minds of other human beings. Ego Tunnels develop in a social environment, and the nature of this environment determines to what extent one Ego Tunnel can resonate with other Ego Tunnels.

So far, we have been concerned only with how the world and the self appear in the tunnel created by the brain. But what about *other* selves? How can other agents with other goals, other thinkers of thoughts, other feeling selves, become parts of one's own inner reality? We can also express this question in philosophical terms. At the beginning of this book, we asked how a first-person perspective can emerge in the brain. The answer was that it does so through the creation of the Ego Tunnel. Now we can ask, What about the *second-person perspective*? Or the "we," the first-person-plural perspective? How does the conscious brain manage to get from the "I" to the "you" and the "we"? The thoughts, goals, feelings, and needs of other living beings in our environment constitute part of our own reality; therefore, it is vital to understand how our brains were able to represent and create not just the inward perspective of the Ego Tunnel but also a world containing multiple Egos and multiple perspectives. Perhaps we will discover that large parts of the first-person perspective did not simply emerge in the brain but were in part causally enabled by the social context we all found ourselves in from the very beginning.

The self-model theory holds that certain new layers of consciousness, unique to the self-model of *Homo sapiens,* made the transition from biological to cultural evolution possible. This process started on an unconscious, automatic level in our brains, and its roots reach far down into the animal kingdom. There is an evolutionary continuity to such high-level social phenomena as the unique human capacity for consciously acknowledging others as rational subjects and moral persons. In chapter 2, I pointed out that in the history of ideas, the concept of "consciousness" was intimately related to possession of a "conscience"—the higher-order ability to assess the moral value of your lower-order mental states or your behavior. What kind of self-model do you need in order to become such a moral agent? The answer could have to do with the progression from a mental representation of the first-person-singular perspective to that of the first-person plural, along with the ability to represent mentally what the benefits (or risks) of a particular action would be for the collective as a whole. You become a moral agent by taking the coherence and stability of your group into account. In this way, the evolution of morals may have had a lot to do with an organism's ability to distance itself mentally from a representation of its individual interests and consciously and explicitly to represent principles of group selection, even if this involved self-damaging behavior. Recall that the beautiful early philosophical theories of consciousness-as-conscience rested on installing an ideal observer in your mind. I believe the human self-model was successful because it installed *your social group* as an ideal observer in your mind, and to a much stronger degree than was the case in any other primate brain. This created a dense causal linkage between global group-control and global self-control—a new kind of ownership, as it were.

Investigators of these phenomena will have to look at chimpanzees and macaques, at swarms of fish and flocks of birds, and maybe even at ant colonies. They will also have to look at the way infants imitate their parents' facial expressions. Intersubjectivity started deep down in the realms of biological behavior coordination, in the motor regions of the brain and the unconscious layers of the Ego. Intersubjectivity is anchored in intercorporality.

SOCIAL NEUROSCIENCE:
CANONICAL NEURONS AND MIRROR NEURONS

Sociological and biological approaches to human consciousness have traditionally been treated as antagonistic to each other, or at least mutually exclusive. But today, in the new discipline of social neuroscience, the assumption is that a multilevel integrative analysis may be required and that a common scientific language, grounded in the structure and function of the brain, can contribute to it. The self-model theory is an attempt to develop exactly this type of language.

It has been known since the 1980s that there is a particularly interesting class of neurons in an area called F5 in the ventral premotor region of the monkey brain. These neurons are part of the unconscious self-model; they code body movements in a highly abstract way. Giacomo Rizzolatti, a professor of human physiology at the University of Parma and a pioneer in this exciting field of research, uses the concept of a "motor vocabulary" that consists of complex inner images of actions as a whole. Words in the monkey's motor vocabulary might be "reach," "grasp," "tear," or "hold." The interesting aspect of this discovery is that there is a specific part of the brain that describes the monkey's—and our own—actions in a holistic manner. This description includes the goals of the actions and the temporal pattern in which the actions unfold. The actions are portrayed as *relations* between an agent and the target object (a piece of fruit, say) of his action.[1]

Now we know that human beings, too, possess something similar. From a neurocomputational perspective, this system in our brains makes sense: By developing an inner vocabulary for possible actions, we reduce the immense space of possibilities to a small number of stereotypical body movements. This allows us, for instance, to perform the same grasping movement in widely differing situations (recall the Alien Hand syndrome of chapter 4).

One of the most fascinating features of these so-called *canonical neurons* is that they also respond to the visual perception of objects in our environment. Our brain does not simply register a chair, a teacup, an apple; it immediately represents the seen object as *what I could do with it*—as an affordance, a set of possible behaviors. This is something I could sit on, this is something I could hold in my hands, this is something I could throw. While we're seeing an object, we are also unconsciously swimming in a sea of possible behaviors. As it turns out, the traditional philosophical distinction between perception and action is an artificial one. In reality, our brains employ a common coding: Everything we perceive is automatically portrayed as a factor in a possible interaction between ourselves and the world. A new medium is created, blending action and perception into a novel, unified representational format. The second fascinating discovery about canonical neurons is that you also use them for self-representation. The motor vocabulary is part of the unconscious self-model, because it describes the goal-directed movements of one's body. The unconscious precursors of the phenomenal Ego in our brain thus play an essential and central role in our perception of the world around us.

In the 1990s, researchers discovered another group of neurons. Also a part of area F5, they fire not just when monkeys perform object-directed actions, such as grasping a peanut, but also when they observe others performing the same type of action. Because these neurons respond to actions performed by others, they are termed *mirror neurons*. They are activated when another agent is observed using objects in a purposeful way. Thus, we are matching the bodily behaviors we observe in others with our own internal motor vocabulary. This action/observation matching system helps us understand something we could never understand using our sensory organs alone—that other beings in our environment pursue goals. We use our own unconscious self-model to put ourselves in the shoes of others, as it were. We use our own "motor ideas" to understand someone else's actions by directly mapping them onto our own inner repertoire, by automatically triggering an inner image of what *our* goal would be if our body also moved that way.[2] The conscious experience of understanding another human being, the subjective feeling that pops up in the Ego Tunnel when we intuitively grasp what others' goals are and what is going on in their minds, is the direct result of these unconscious processes.[3]

The conscious self is thus not only a window into the internal workings of one's own Ego but also a window into the social world. It is a two-way window: It elevates to the level of global availability the unconscious and automatic processes that organisms constantly use to represent one another's behavior. This is how these processes become part of the Ego Tunnel, an element of our subjective reality. They lead to an enormous expansion and enrichment of our inner simulation of the world. As soon as our brains are able to represent not only events but also actions—that is, goal-directed events caused by other beings—we are not alone anymore. Others exist, with minds of their own. The fact that more than one Ego Tunnel might exist in the world is now reflected in our own tunnel. We can develop our conscious action-ontology, and we can put it to use by sharing it with others.[4]

A considerable body of evidence using a variety of neuroimaging techniques shows that the mirror-neuron system exists not just in monkeys but in humans as well. However, it appears that the system in humans is much more generalized and does not depend on concrete effector-object interactions; consequently, it can represent a much greater variety of actions than it does in monkeys. In particular, researchers have now discovered mirror-neuron systems that seem to achieve similar effects for emotions and for pain and other bodily sensations. When human test subjects are shown pictures of sad faces, for example, they subsequently tend to rate themselves as sadder than they were before—and after being shown happy faces they tend to rate themselves as happier. Converging empirical data show that when we observe other human beings expressing emotions, we simulate them with the help of the same neural networks that are active when we feel or express these emotions ourselves.[5] For instance, certain regions in the insular cortex are activated when subjects are exposed to a disgusting smell, and the same regions are active when we see an expression of disgust on another person's face. A common representation of the emotional state of disgust is activated in our brains whether we experience it ourselves or observe it in another individual. Parallel observations in the amygdala have been made for fear.[6] It is interesting to note that our ability to recognize a particular feeling in another human being can be weakened or switched off by blocking the relevant parts of the mirror-neuron system. It is believed, for example, that certain areas in the ventral striatum of the basal ganglia are necessary in recognizing anger; patients with damage to this area show impairment in identifying aggression signals emitted by others. If these areas are blocked pharmacologically (by interfering with dopamine metabolism), subjects can recognize other emotions but can no longer recognize anger.[7] Similar observations have been made for pain. Recent fMRI (functional magnetic resonance imaging) experiments show that areas in the anterior cingulate cortex and the interior insular cortex are active when we experience pain but also when we observe someone else experiencing pain.[8] Interestingly, only the emotional part of the pain system is activated; the part associated with the purely sensory aspect of pain is not. This makes perfect sense, because the sensory aspect is exactly what we cannot share with anyone else: We cannot share the cutting, throbbing, or burning sensory quality of pain, but we can feel empathy with regard to the emotions it causes.

Other neuroimaging experiments have demonstrated that a similar principle exists for other bodily sensations. Certain higher levels of the somatosensory cortex are activated both when subjects observe others being touched and when they are touched themselves. Again, the immediate sensory quality associated with the activation of the primary somatosensory cortex cannot be shared, but a higher level in the bodily Ego is active regardless of whether we are being touched or just observing someone being touched. There seems to be an underlying principle uniting these new empirical discoveries: Certain layers of our self-model function as a bridge to the social domain, because they can directly map abstract inner descriptions of what is going on in ourselves onto those of what goes on in other people.

Of course, intersubjectivity is not only about the body and emotions. Thinking plays a role as well. Reason-based forms of empathy appear to involve yet other parts of the brain—specifically, the ventromedial prefrontal cortex. Still, the discovery of mirror neurons helps us to understand that empathy is a natural phenomenon, acquired step by step in the course of our biological evolution. First, we developed the self-model, because we had to integrate our sensory perceptions with our bodily behavior. Then this self-model became conscious, and the phenomenal self-model was born into the Ego Tunnel, allowing us to achieve global control of our bodies in a much more selective and flexible manner. This was the step from being an embodied natural system that has and uses an internal image of itself as a whole to a system that, in addition, consciously experiences this fact.[9] The next evolutionary step was what Vittorio Gallese, Rizzolatti's colleague at Parma and one of the leading researchers in the field, has called *embodied simulation*.[10] In order to understand the feelings and goals of other human beings, we use our own body-model in the brain to simulate them.

As recent neuroscientific findings show, this process also cuts across the border between the unconscious and the conscious. A considerable part of this constant mirroring activity happens outside the Ego Tunnel, and thus we have no subjective experience of it. But from time to time, when we deliberately attend to other people or analyze social situations, the conscious self-model is involved as well; in particular, as noted, we can somehow directly comprehend, almost perceive, what somebody else is up to. Often, we "just know" what the purpose of the other person's action is and what his likely emotional state is.[11] We use the same internal resources that make us aware of our own goal states to discover automatically that others are goal-directed entities themselves and not just other moving objects. We can experience them as Egos because we experience ourselves as Egos. Whenever successful social understanding and empathy are achieved, we share a common representation: of one and the same goal state in two different Ego Tunnels. Social cognition has now become tractable to empirical neuroscience on the level of single-cell recordings—showing us not only how Ego Tunnels started to resonate with each other but also how complex cooperation and communication between self-conscious organisms were able to evolve and lay the foundations for cultural evolution.

My idea is that social cognition rests on what is sometimes called an *exaptation*. Adaptation led to an integrated body-model in the brain and to the phenomenal self-model. Then the existing neural circuitry was "exapted" for another form of intelligence: It suddenly proved useful in tackling a different set of problems. This process began with low-order motor resonance; then, second- and third-order embodiment[12] led to embodied simulation as a brand-new tool in developing social intelligence. Like everything else in evolution, this process was driven by chance. There was no purpose behind it, but it eventually led us where we are today—to the formation of intelligent, scientific communities peopled by conscious agents trying to understand this very process itself.

The new emerging general picture is inspiring: We are all constantly swimming in an unconscious sea of intercorporality, permanently mirroring one another with the aid of various unconscious components and precursors of the phenomenal Ego. Long before conscious, high-level social understanding arrived on the scene, and long before language evolved and philosophers developed complicated theories about what it takes for one human being to acknowledge another as a person and a rational individual, we were already bathed in the waters of implicit, bodily intersubjectivity. Few great social philosophers of the past would have thought that social understanding had anything to do with the premotor cortex, and that "motor ideas" would play such a central role in the emergence of social understanding. Who could have expected that shared thought would depend upon shared "motor representations"? Or that the functional aspects of the human self-model that are necessary for the development of social consciousness are nonconceptual, prerational, and pretheoretical? The first inklings of these ideas came at the end of the nineteenth and the first half of the twentieth century, when there were numerous attempts in experimental psychology to better understand so-called ideomotor phenomena.[13] Philosopher Theodor Lipps wrote about *Einfühlung* (empathy) in 1903—that is, the ability, as he put it, to "feel yourself in an object." He had already spoken of "inner imitation" and of "organic feelings." For him, objects of empathy could be not only the movements or postures we perceive in other human beings but also objects of art, architecture, and even visual illusions. He held that aesthetic pleasure was "objectified"—that is, "the object is ego and thereby the ego object." [14] Social psychologists began talking about concepts such as "virtual body movements" and "motor mimicry" or "motor infection" decades ago.

From a philosophical perspective, the discovery of mirror neurons is exciting because it gave us an idea of how motor primitives could have been used as semantic primitives: that is, how *meaning* could be communicated between agents. Thanks to our mirror neurons, we can consciously experience another human being's movements as meaningful. Perhaps the evolutionary precursor of language was not animal calls but gestural communication.[15] The transmission of meaning may initially have grown out of the unconscious bodily self-model and out of motor agency, based, in our primate ancestors, on elementary gesturing. Sounds may only later have been associated with gestures, perhaps with facial gestures—such as scowling, wincing, or grinning—that already carried meaning. Still today, the silent observation of another human being grasping an object is immediately understood, because, without symbols or thought in between, it evokes the same motor representation in the parieto-frontal mirror system of our own brain. As Professor Rizzolatti and Dr. Maddalena Fabbri Destro from the Department of Neuroscience at the University of Parma put it: "[T]he mirror mechanism solved, at an initial stage of language evolution, two fundamental communication problems: parity and direct comprehension. Thanks to the mirror neurons, what counted for the sender of the message also counted for the receiver. No arbitrary symbols were required. The comprehension was inherent in the neural organization of the two individuals."[16]

Such ideas give a new and rich meaning not only to the concepts of "grasping" and "mentally grasping the intention of another human being," but, more important, also to the concept of grasping a *concept*—the essence of human thought itself. It may have to do with simulating hand movements in your mind but in a much more abstract manner. Humankind has apparently known this for centuries, intuitively: "Concept" comes from the Latin *conceptum*, meaning "a thing conceived," which, like our modern "to conceive of something," is rooted in the Latin verb *concipere*, "to take in and hold." As early as 1340, a second meaning of the term had appeared: "taking into your mind." Surprisingly, there is a representation of the human hand in Broca's area, a section of the human brain involved in language processing, speech or sign production, and comprehension. A number of studies have shown that hand/arm gestures and movements of the mouth are linked through a common neural substrate. For example, grasping movements influence pronunciation—and not only when they are executed but also when they are observed. It has also been demonstrated that hand gestures and mouth gestures are directly linked in humans, and the oro-laryngeal movement patterns we create in order to produce speech are a part of this link.

Broca's area is also a marker for the development of language in human evolution, so it is intriguing to see that it also contains a motor representation of hand movements; here may be a part of the bridge that led from the "body semantics" of gestures and the bodily self-model to linguistic semantics, associated with sounds, speech production, and abstract meaning expressed in our cognitive self-model, the thinking self. Broca's area is present in fossils of *Homo habilis,* whereas the presumed precursors of these early hominids lacked it. Thus the mirror mechanism is conceivably the basic mechanism from which language evolved. By providing motor copies of observed actions, it allowed us to extract the action goals from the minds of other human beings—and later to send abstract meaning from one Ego Tunnel to the next.

The mirror-neuron story is attractive not only because it bridges neuroscience and the humanities but also because it illuminates a host of simpler social phenomena. Have you ever observed how infectious a yawn is? Have you ever caught yourself starting to laugh out loud with others, even though you didn't really understand the joke? The mirror-neuron story gives us an idea of how groups of animals—fish schools, flocks of birds—can coordinate their behavior with great speed and accuracy; they are linked through something one might call a low-level resonance mechanism. Mirror neurons can help us understand why parents spontaneously open their mouths while feeding their babies, what happens during a mass panic, and why it is sometimes hard to break away from the herd and be a hero. Neuroscience contributes to the image of humankind: We are all connected in an intersubjective space of meaning—what Vittorio Gallese calls a "shared manifold."[17]

CHAPTER SIX APPENDIX
THE SHARED MANIFOLD:
A CONVERSATION WITH
VITTORIO GALLESE

 **Vittorio Gallese** is professor of human physiology in the Department of Neurosciences of the University of Parma, Italy. As a cognitive neuroscientist, he focuses his research interests on the relationship between the sensory-motor system and cognition in primates, both human and nonhuman, using a variety of neurophysiological and neuroimaging techniques. Among his major contributions is the discovery, with his colleagues in Parma, of mirror neurons and the elaboration of a theoretical model of the basic aspects of social cognition. He is developing an interdisciplinary approach to the understanding of intersubjectivity and social cognition, in collaboration with psychologists, psycholinguists, and philosophers. In 2002, he was the George Miller visiting professor at the University of California at Berkeley. In 2007 he received the Grawemeyer Award for Psychology for the discovery of mirror neurons. He has published more than seventy papers in international journals and is coeditor (with Maxim I. Stamenov) of *Mirror Neurons and the Evolution of Brain and Language* (2002).

Metzinger: Vittorio, what exactly do you mean by the shared manifold hypothesis. What is a shared manifold?

Gallese: The question I started out with is the following: How can we explain the ease with which we normally understand what is at stake when we interact with other people?

I used this term to characterize what happens when we witness the actions of others, or their overt behavior expressing the sensations and emotions they experience. Basically, it describes our capacity for direct and implicit access to the experiential world of the other. I think the concept of empathy should be extended in order to accommodate and account for all different aspects of expressive behavior enabling us to establish a meaningful link with others. This enlarged notion of empathy is captured by the term "shared manifold." It opens up the possibility of giving a unified account of important aspects and possible levels of description of intersubjectivity. I tried on purpose not to employ the term "empathy," because it systematically induces misunderstandings, mainly because of its different connotations in different contexts. The shared manifold can be described at three different levels: a phenomenological level, a functional level, and a subpersonal level.

The phenomenological level is the one responsible for the sense of similarity—of being part of a larger social community of persons like us—that we experience anytime we encounter others. When confronting the intentional behavior of others, we experience a specific phenomenal state of intentional attunement. This phenomenal state generates the peculiar quality of familiarity with other individuals, produced by the collapse of the others' intentions into those of the observer. This seems to be one important component of what being empathic is about.

The functional level can be characterized in terms of embodied simulations of the actions we see or of the emotions and sensations whose expression we observe in others.

The subpersonal level is instantiated as the activity of a series of mirroring neural circuits. The activity of these mirror neural circuits is, in turn, tightly coupled with multilevel changes within body-states.

We have seen that mirror neurons instantiate a multimodal shared space for actions and intentions. Recent data show that analogous neural networks are at work to generate multimodal emotional and sensitive "we-centric" shared spaces. To put it in simpler words, every time we relate to other people, we automatically inhabit a we-centric space, within which we exploit a series of implicit certainties about the other. This implicit knowledge enables us to understand in a direct way what the other person is doing, why he or she is doing it, and how he or she feels about a specific situation.

Metzinger: You also speak of "embodied simulation." What exactly does that mean? Is there also something like "disembodied simulation"?

Gallese: The notion of simulation is employed in many different domains, often with different, not necessarily overlapping, meanings. Simulation is a functional process that possesses a certain representational content, typically focusing on possible states of its target object. In philosophy of mind, the notion of simulation has been used by the proponents of the "Simulation Theory of Mind-Reading" to characterize the pretend state adopted by the attributer in order to understand another person's behavior. Basically, we use our mind to put ourselves into the mental shoes of other human beings.

I qualify simulation as embodied in order to characterize it as a mandatory, automatic, nonconscious, prerational, nonintrospectionist process. A direct form of experiential understanding of others, intentional attunement, is achieved by the activation of shared neural systems underpinning what others do and feel and what we do and feel. This modeling mechanism *is* embodied simulation. Parallel to the detached sensory description of the observed social stimuli, internal representations of the body-states associated with actions, emotions, and sensations are evoked in the observer, *as if* he or she were performing a similar action or experiencing a similar emotion or sensation. Mirror-neuron systems are likely the neural correlate of this mechanism. By means of a shared neural state realized in two different physical bodies, the "objectual other" becomes another self. Defective intentional attunement, caused by a lack of embodied simulation, might explain some of the social impairments of autistic individuals.

I should add that—in contrast to what many cognitive scientists think—social cognition is not only social metacognition, that is, explicitly thinking about the contents of someone else's mind by means of abstract representations. We can certainly explain the behavior of others by using our complex and sophisticated mentalizing ability. My point is that most of the time in our daily social interactions, we do not need to do this. We have a much more direct access to the experiential world of the other. This dimension of social cognition is embodied, in that it mediates between our multimodal experiential knowledge of our own lived body and the way we experience others. I therefore call simulation "embodied"—not only because it is realized in the brain but also because it uses a preexisting body-model in the brain and therefore involves a nonpropositional form of self-representation that also allows us to experience what others are experiencing.

Metzinger: Vittorio, according to our best current theories, what exactly is the difference between social cognition in monkeys or chimps and social cognition in human beings?

Gallese: The traditional view in the cognitive sciences holds that humans are able to understand the behavior of others in terms of their own mental states—intentions, beliefs, and desires—by exploiting what is commonly called folk psychology. The capacity for attributing mental states to others has been defined as "theory of mind." A common trend on this issue has been to emphasize that nonhuman primates, apes included, do not rely on mentally based accounts of one another's behavior.

This view prefigures a sharp distinction between all nonhuman species, which are confined to behavior reading, and our species, which makes use of a different level of explanation—mind-reading. However, it is by no means obvious that behavior-reading and mind-reading constitute two autonomous realms. As I said before, in our social transactions we seldom engage in explicit interpretive acts.

Most of the time, our understanding of social situations is immediate, automatic, almost reflex-like. Therefore, I think it is preposterous to claim that our capacity for reflecting on the real intentions determining the behavior of others is all there is to social cognition. It is even less obvious that in understanding the intentions of others, we employ a cognitive strategy totally unrelated to predicting the consequences of their observed behavior.

The use of the belief/desire propositional attitudes of folk psychology in social transactions is probably overstated. As emphasized by Jerry [Jerome S.] Bruner, "When things are as they should be, the narratives of Folk Psychology are unnecessary."[18]

Furthermore, recent evidence shows that fifteen-month-old infants recognize false beliefs. These results suggest that typical aspects of mind-reading, such as the attribution of false beliefs to others, can be explained on the basis of low-level mechanisms that develop well before full-blown linguistic competence.

The all-or-nothing approach to social cognition of mainstream cognitive science—its search for a mental Rubicon, the wider the better—is strongly arguable. When trying to understand our social-cognitive abilities, we should not forget that they are the result of a long evolutionary process. It is therefore possible that apparently different cognitive strategies are underpinned by similar functional mechanisms, which in the course of evolution acquire increasing complexity and are exapted to sustain cognitive skills newly emerged out of the pressure exerted by changed social and/or environmental constraints. Before drawing any firm conclusion about the mentalizing abilities of nonhuman species, methodological issues related to species-specific spontaneous abilities and environments should be carefully scrutinized.

A fruitful alternative strategy I fully endorse is that of framing the issue of the investigation of the neural bases of social cognition within an evolutionary perspective. The evolution of this cognitive trait seems to be related to the necessity of dealing with social complexities that arose when group-living individuals had to compete for scarce and patchily distributed resources.

Cognitive neuroscience has started to unveil, both in monkeys and in humans, the neural mechanisms at the basis of anticipating and understanding the actions of others and the basic intentions promoting them—the mirror-neuron system for action. The results of this ongoing research can shed light on the evolution of social cognition. The empirical data on mirror neurons in monkeys and on mirroring circuits in the human brain suggest that some of the typically human, sophisticated mentalizing skills—such as ascribing intentions to others—might be the outcome of a continuous evolutionary process, whose antecedent stages can be traced to the mirror mapping system of macaque monkeys.

Thus, as you are asking, what makes humans different? Language certainly plays a key role. But in a sense this answer begs the question, because then we must explain why we have language and other animals do not. At present, we can only make hypotheses about the relevant neural mechanisms underpinning the mentalizing abilities of humans, which are still poorly understood from a functional point of view.

One distinctive feature of our mentalizing abilities is our capacity for entertaining potentially infinite orders of intentionality: "I know that you know that I know . . ." and so on. One important difference between humans and monkeys could be the higher level of recursion attained—among other neural systems—by the mirror-neuron system for actions in our species. A similar proposal has recently been put forward in relation to the faculty of language, another human faculty characterized by recursion and generativity. Our species is capable of mastering hierarchically complex phrase-structure grammars, while nonhuman primates are confined to the use of much simpler finite-state grammars. A quantitative difference in computational power and degree of recursion could produce a qualitative leap forward in social cognition.

Metzinger: Can you speculate about the role of mirror neurons in the transition from biological to cultural evolution?

Gallese: A possibility is that mirror neurons and the embodied simulation mechanisms they underpin might be crucial for learning how to use the cognitive tools of folk psychology. This typically occurs when children are repeatedly exposed to the narration of stories. In fact, embodied simulation is certainly at play during language processing. But certainly the aspect of human culture that is more likely to benefit from mirror neurons is the domain of imitation, the domain of our incredibly pervasive mimetic skills. If it is true that ours is basically a mimetic culture, then mirror neurons, which are deeply involved in imitation and imitative learning, certainly are one important and basic ingredient of this crucial cultural transition. And indeed there is plenty of evidence that when we imitate simple motor acts, such as lifting a finger, or learn complex motor sequences, as when learning to play chords on a guitar, we do this by employing our mirror neurons. But instead of drawing a line between species like ours, who are fully competent in imitation, and other species, where this capacity is at best only emerging—again, we are dealing here with the anthropocentric dichotomies so appealing to many of my colleagues—we should concentrate on understanding why mimetic skills are so important for the cultural evolution of our species. And to answer this question, we must place the issue of mimesis in the larger context of our peculiar social cognition, in which the period of parental care is much longer than in all other species. There is a clear-cut relationship between the prolonged dependency of infants on their parents and the learning processes that this dependency promotes. The longer the period of infantile dependency, the greater the opportunities to develop complex emotional and cognitive strategies of communication. Increased communication in turn fosters cultural evolution. Given the central role that mirror neurons seem to play in establishing meaningful bonds among individuals, their connection to cultural evolution seems very plausible.

For most of history, the culture of our species has been an oral culture, where the transmission of knowledge from one generation to the next had to rely on direct personal contact between the transmitter of cultural content and the recipient of the cultural transmission. As pointed out by scholars like [Walter J.] Ong and

[Eric A.] Havelock, for millennia cultural transmission had to rely on the same cognitive apparatus we still exploit in our interpersonal transactions—that is, our ability to identify and empathize with others. Again, I think that if we look at cultural evolution from this particular perspective, the role of mirror neurons appears to be central. At present, we are witnessing a cultural paradigm shift. The impact of new technologies, such as cinema, television, and more recently the Internet, with its massive introduction of multimediality, is drastically changing the way in which we communicate knowledge. The mediated, objective status of culture as transported by written texts like books is progressively being supplemented with a more direct access to the same contents by means of the new media of cultural fruition. This media revolution will most likely introduce cognitive changes, and I suspect that mirror neurons will again be involved.

Metzinger: In the field of social cognitive neuroscience, what do you consider to be the most burning and urgent questions for the future, and in which direction is the field moving?

Gallese: The first point I would like to make is a methodological one. I think we should definitely try to focus more strongly on the nature of the subjects of our investigations. Most of what we know about the neural aspects of social cognition—with few exceptions pertaining to the study of language—derives from brain-imaging studies carried out on Western-world psychology students! Even with present technologies, we could do a lot better than this. It is an open question whether cognitive traits and the neural mechanisms underpinning them are universal or, at least to a degree, the product of a particular social environment and cultural education. To answer this question, we need an ethno-neuroscience.

Second, even within the average sample of subjects normally studied by social cognitive neuroscientists, we do not know—or at best know very little—to what extent the results correlate with specific personality traits, gender, professional expertise, and the like. In sum, we should move from the characterization of an unrealistic "average social brain" to a much more fine-grained characterization.

A third issue I would like to see addressed more specifically in the near future is the role played by embodied mechanisms in semantic and syntactical aspects of language. Let me be clear about this. Even though I spent a considerable part of my scientific career investigating prelinguistic mechanisms in social cognition, I do not think you can avoid language if the ultimate goal is to understand what social cognition really is. All our folk psychology is language-based. How does this square with the embodied approach to social cognition? To me, this is a burning question.

A fourth important point pertains to the phenomenological aspects of social cognition. I think we should try to design studies in which a correlation can be drawn between particular patterns of brain activation and specific qualitative subjective experiences. Single case studies are now possible with high-resolution brain imaging. I am fully aware that dealing with subjective states is a tricky issue, from which empirical science so far has tried to stay clear, for many good reasons. But in principle it should be possible to carefully design well-suited and well-controlled experimental paradigms to crack the boundaries of subjective phenomenal states.

Metzinger: Vittorio, you have repeatedly cornered me with pressing questions about Edmund Husserl, Maurice Merleau-Ponty, and Edith Stein. Why are you so interested in philosophy, and what kind of philosophy would you like to see in the future? What relevant contributions from the humanities are you expecting?

Gallese: Scientists who believe that their discipline will progressively eliminate *all* philosophical problems are simply fooling themselves. What science can contribute to is the elimination of *false* philosophical problems. But this is a totally different issue.

If our scientific goal is to understand what it means to be human, we need philosophy to clarify what issues are at stake, what problems need to be solved, what is epistemologically sound and what is not. Cognitive neuroscience and philosophy of mind deal with the same problems but use different approaches and different levels of descriptions. Very often, we use different words to speak about the same things. I think all cognitive neuroscientists should take classes in philosophy. Similarly, philosophers—at least, philosophers of mind—should learn a lot more about the brain and how it works. We need to talk to one another much more than we are doing now. How can you possibly investigate social cognition without knowing what an intention is, or without understanding the concept of second-order intentionality? Similarly, how can you possibly stick to a philosophical theory of cognition if it is patently falsified by the available empirical evidence? There is another aspect for which I think philosophy may be helpful. Our scientific bravado sometimes makes us think we are the first to have thought about something. Most of the time, this is not true!

As I said, philosophy should listen more carefully to the results of cognitive neuroscience. But things are changing rapidly. The current situation is much better than it was ten years ago. There are more and more chances for multidisciplinary exchanges between our disciplines. One of my PhD students, currently involved in neurophysiological experiments, has a degree in philosophy.

Broadening these considerations to the overall field of the humanities, I think incredibly fruitful contributions can result from a dialogue with anthropology, aesthetics, and literary and film studies. As I said before, a mature social cognitive neuroscience can't limit itself to scanning brains in a lab. It must be open to the contributions from all these disciplines. I am rather optimistic. I see a future of ever-growing and stimulating dialogue between cognitive neuroscience and the humanities.

PART THREE | **THE CONSCIOUSNESS REVOLUTION**

:✦:

SEVEN

ARTIFICIAL EGO MACHINES

From this point on, let us call any system capable of generating a conscious self an *Ego Machine*. An Ego Machine does not have to be a living thing; it can be anything that possesses a conscious self-model. It is certainly conceivable that someday we will be able to construct artificial agents. These will be self-sustaining systems. Their self-models might even allow them to use tools in an intelligent manner. If a monkey's arm can be replaced by a robot arm and a monkey's brain can learn to directly control a robot arm with the help of a brain-machine interface, it should also be possible to replace the entire monkey. Why should a robot not be able to experience the rubber-hand illusion? Or have a lucid dream? If the system has a body model, full-body illusions and out-of-body experiences are clearly also possible.

In thinking about artificial intelligence and artificial consciousness, many people assume there are only two kinds of information-processing systems: artificial ones and natural ones. This is false. In philosophers' jargon, the conceptual distinction between natural and artificial systems is neither *exhaustive* nor *exclusive:* that is, there could be intelligent and/or conscious systems that belong in neither category. With regard to another old-fashioned distinction—software versus hardware—we already have systems using biological hardware that can be controlled by artificial (that is, man-made) software, and we have artificial hardware that runs naturally evolved software.

Hybrid biorobots are an example of the first category. Hybrid biorobotics is a new discipline that uses naturally evolved hardware and does not bother with trying to re-create something that has already been optimized by nature over millions of years. As we reach the limitations of artificial computer chips, we may increasingly use organic, genetically engineered hardware for the robots and artificial agents we construct.

An example of the second category is the use of software patterned on neural nets to run in artificial hardware. Some of these attempts are even using the neural nets themselves; for instance, cyberneticists at the University of Reading (U.K.) are controlling a robot by means of a network of some three hundred thousand rat neurons.[1] Other examples are classic artificial neural networks for language acquisition or those used

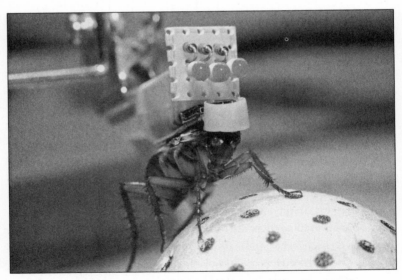

Figure 16: RoboRoach. Controlling the movements of cockroaches with surgically implanted microrobotic backpacks. The roach's "backpack" contains a receiver that converts the signals from a remote control into electrical stimuli that are applied to the base of the roach's antennae. This allows the operator to get the roach to stop, go forward, back up, or turn left and right on command.

by consciousness researchers such as Axel Cleeremans at the Cognitive Science Research Unit at Université Libre de Bruxelles in Belgium to model the metarepresentational structure of consciousness and what he calls its "computational correlates."[2] The latter two are biomorphic and only semiartificial information-processing systems, because their basic functional architecture is stolen from nature and uses processing patterns that developed in the course of biological evolution. They create "higher-order" states; however, these are entirely subpersonal.

We may soon have a functionalist theory of consciousness, but this doesn't mean we will also be able to implement the functions this theory describes on a nonbiological carrier system. Artificial consciousness is not so much a theoretical problem in philosophy of mind as a technological challenge; the devil is in the details. The real problem lies in developing a non-neural kind of hardware with the right causal powers: Even a simplistic, minimal form of "synthetic phenomenology" may be hard to achieve—and for purely technical reasons.

The first self-modeling machines have already appeared. Researchers in the field of artificial life began simulating the evolutionary process long ago, but now we have the academic discipline of "evolutionary robotics." Josh Bongard, of the Department of Computer Science at the University of Vermont, and his colleagues Victor Zykov and Hod Lipson have created an artificial starfish that gradually develops an explicit internal self-model.[3] Their four-legged machine uses actuation-sensation relationships to infer indirectly its own structure and then uses this self-model to generate forward locomotion. When part of its leg is removed, the machine adapts its self-model and generates alternative gaits—it learns to limp. Unlike the phantom-limb patients discussed in chapter 4, it can restructure its body representation following the loss of a limb; thus, in a sense, it can learn. As its creators put it, it can "autonomously recover its own topology with little prior knowledge," by constantly optimizing the parameters of its resulting self-model. The starfish not only synthesizes an internal self-model but also uses it to generate intelligent behavior.

Self-models can be unconscious, they can evolve, and they can be created in machines that mimic the process of biological evolution. In

Figure 17a. Starfish, a four-legged robot that walks by using an internal self-model it has developed and which it continuously improves. If it loses a limb, it can adapt its internal self-model.[5]

sum, we already have systems that are neither exclusively natural nor exclusively artificial. Let us call such systems *postbiotic*. The likely possibility is that conscious selfhood will first be realized in postbiotic Ego Machines.

HOW TO BUILD AN ARTIFICIAL CONSCIOUS SUBJECT AND WHY WE SHOULDN'T DO IT

Under what conditions would we be justified in assuming that a given postbiotic system has conscious experience? Or that it also possesses a conscious *self* and a genuine consciously experienced *first-person perspective?* What turns an information-processing system into a subject of experience? We can nicely sum up these questions by asking a simpler and more provocative one: What would it take to build an artificial Ego Machine?

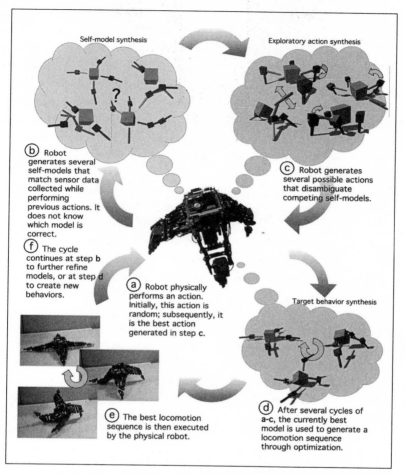

Figure 17b: The robot continuously cycles through action execution. (A and B) Self-model synthesis. The robot physically performs an action (A). Initially, this action is random; later, it is the best action found in (C). The robot then generates several self-models to match sensor data collected while performing previous actions (B). It does not know which model is correct. (C) Exploratory action synthesis. The robot generates several possible actions that disambiguate competing self-models. (D) Target behavior synthesis. After several cycles of (A) to (C), the best current model is used to generate locomotion sequences through optimization. (E) The best locomotion sequence is executed by the physical device. (F)[4]

Being conscious means that a particular set of facts is available to you: that is, all those facts related to your *living in a single world.* Therefore, any machine exhibiting conscious experience needs an integrated and dynamical world-model. I discussed this point in chapter 2, where I pointed out that every conscious system needs a unified inner representation of the world and that the information integrated by this representation must be simultaneously available for a multitude of processing mechanisms. This phenomenological insight is so simple that it has frequently been overlooked: Conscious systems are systems operating on globally available information with the help of a single internal model of reality. There are, in principle, no obstacles to endowing a machine with such an integrated inner image of the world and one that can be continuously updated.

Another lesson from the beginning of this book was that, in its very essence, consciousness is *the presence of a world.* In order for a world to appear to it, an artificial Ego Machine needs two further functional properties. The first consists of organizing its internal information flow in a way that generates a psychological moment, an experiential Now. This mechanism will pick out individual events in the continuous flow of the physical world and depict them as contemporaneous (even if they are not), ordered, and flowing in one direction successively, like a mental string of pearls. Some of these pearls must form larger gestalts, which can be portrayed as the experiential content of a single moment, a lived Now. The second property must ensure that these internal structures cannot be recognized by the artificial conscious system as internally constructed images. They must be transparent. At this stage, a world would appear to the artificial system. The activation of a unified, coherent model of reality within an internally generated window of presence, when neither can be recognized as a model, is the appearance of a world. In sum, the appearance of a world is consciousness.

But the decisive step to an Ego Machine is the next one. If a system can integrate an equally transparent internal image of itself into this phenomenal reality, then it will appear to itself. It will become an Ego and a naive realist about whatever its self-model says it is. The phenomenal property of selfhood will be exemplified in the artificial system, and it will appear to itself not only as *being someone* but also as *being there*. It will believe in itself.

Note that this transition turns the artificial system into an object of moral concern: It is now potentially able to *suffer*. Pain, negative emotions, and other internal states portraying parts of reality as undesirable can act as causes of suffering only if they are consciously owned. A system that does not appear to itself cannot suffer, because it has no sense of ownership. A system in which the lights are on but nobody is home would not be an object of ethical considerations; if it has a minimally conscious world model but no self-model, then we can pull the plug at any time. But an Ego Machine can suffer, because it integrates pain signals, states of emotional distress, or negative thoughts into its transparent self-model and they thus appear as *someone's* pain or negative feelings. This raises an important question of animal ethics: How many of the conscious biological systems on our planet are only phenomenal-reality machines, and how many are actual Ego Machines? How many, that is, are capable of the conscious experience of suffering? Is RoboRoach among them? Or are only mammals, such as the macaques and kittens, sacrificed in consciousness research? Obviously, if this question cannot be decided for epistemological reasons, we must make sure always to err on the side of caution. It is precisely at this stage of development that any theory of the conscious mind becomes relevant for ethics and moral philosophy.

An Ego Machine is also something that possesses a perspective. A strong version should know that it has such a perspective by becoming aware of the fact that it is *directed*. It should be able to develop an inner picture of its dynamical relations to other beings or objects in its environment, even as it perceives and interacts with them. If we do manage to build or evolve this type of system successfully, it will experience itself as interacting with the world—as attending to an apple in its hand, say, or as forming thoughts about the human agents with whom it is communicating. It will experience itself as directed at goal states, which it will represent in its self-model. It will portray the world as containing not just a self but a perceiving, interacting, goal-directed *agent*. It could even have a high-level concept of itself as a subject of knowledge and experience.

Anything that can be represented can be implemented. The steps just sketched describe new forms of what philosophers call *representational content,* and there is no reason this type of content should be restricted to living systems. Alan M. Turing, in his famous 1950 paper "Computing Machinery and Intelligence," made an argument that later was condensed thus by distinguished philosopher Karl Popper in his book *The Self and Its Brain,* which he coauthored with the Nobel Prize–winning neuroscientist Sir John Eccles. Popper wrote: "Specify the way in which you believe a man is superior to a computer and I shall build a computer which refutes your belief. Turing's challenge should not be taken up; for any sufficiently precise specification could be used in principle to programme a computer."[6]

Of course, it is not the self that uses the brain (as Karl Popper would have it)—the brain uses the self-model. But what Popper clearly saw is the dialectic of the artificial Ego Machine: Either you cannot identify what exactly about human consciousness and subjectivity cannot be implemented in an artificial system or, if you can, then it is just a matter of writing an algorithm that can be implemented in software. If you have a precise definition of conciousness and subjectivity in causal terms, you have what philosophers call a *functional analysis.* At this point, the mystery evaporates, and artificial Ego Machines become, in principle, technologically feasible. But *should* we do whatever we're able to do?

Here is a thought experiment, aimed not at epistemology but at ethics. Imagine you are a member of an ethics committee considering scientific grant applications. One says:

> We want to use gene technology to breed mentally retarded human infants. For urgent scientific reasons, we need to generate human babies possessing certain cognitive, emotional, and perceptual deficits. This is an important and innovative research strategy, and it requires the controlled and reproducible investigation of the retarded babies' psychological development after birth. This is not only important for understanding how our own minds work but also has great potential for healing psychiatric diseases. Therefore, we urgently need comprehensive funding.

No doubt you will decide immediately that this idea is not only absurd and tasteless but also dangerous. One imagines that a proposal of this kind would not pass any ethics committee in the democratic world. The point of this thought experiment, however, is to make you aware that the unborn artificial Ego Machines of the future would have no champions on today's ethics committees. The first machines satisfying a minimally sufficient set of conditions for conscious experience and selfhood would find themselves in a situation similar to that of the genetically engineered retarded human infants. Like them, these machines would have all kinds of functional and representational deficits—various disabilities resulting from errors in human engineering. It is safe to assume that their perceptual systems—their artificial eyes, ears, and so on—would not work well in the early stages. They would likely be half-deaf, half-blind, and have all kinds of difficulties in perceiving the world and themselves in it—and if they were true artificial Ego Machines, they would, *ex hypothesi,* also be able to suffer.

If they had a stable bodily self-model, they would be able to feel sensory pain as their own pain. If their postbiotic self-model was directly anchored in the low-level, self-regulatory mechanisms of their hardware— just as our own emotional self-model is anchored in the upper brainstem and the hypothalamus—they would be consciously *feeling* selves. They would experience a loss of homeostatic control as painful, because they had an inbuilt *concern* about their own existence. They would have interests of their own, and they would subjectively experience this fact. They might suffer emotionally in qualitative ways completely alien to us or in degrees of intensity that we, their creators, could not even imagine. In fact, the first generations of such machines would very likely have many negative emotions, reflecting their failures in successful self-regulation because of various hardware deficits and higher-level disturbances. These negative emotions would be conscious and intensely felt, but in many cases we might not be able to understand or even recognize them.

Take the thought experiment a step further. Imagine these postbiotic Ego Machines as possessing a cognitive self-model—as being intelligent thinkers of thoughts. They could then not only conceptually grasp the bizarreness of their existence as mere objects of scientific interest but also could intellectually suffer from knowing that, as such, they lacked the innate "dignity" that seemed so important to their creators. They might well be able to consciously represent the fact of being only second-class sentient citizens, alienated postbiotic selves being used as interchangeable experimental tools. How would it feel to "come to" as an advanced artificial subject, only to discover that even though you possessed a robust sense of selfhood and experienced yourself as a genuine subject, you were only a commodity?

The story of the first artificial Ego Machines, those postbiotic phenomenal selves with no civil rights and no lobby in any ethics committee, nicely illustrates how the capacity for suffering emerges along with the phenomenal Ego; suffering starts in the Ego Tunnel. It also presents a principled argument against the creation of artificial consciousness as a goal of academic research. Albert Camus spoke of the solidarity of all finite beings against death. In the same sense, all sentient beings capable of suffering should constitute a solidarity against suffering. Out of this solidarity, we should refrain from doing anything that could increase the overall amount of suffering and confusion in the universe. While all sorts of theoretical complications arise, we can agree not to *gratuitously* increase the overall amount of suffering in the universe—and creating Ego Machines would very likely do this right from the beginning. We could create suffering postbiotic Ego Machines before having understood which properties of our biological history, bodies, and brains are the roots of our own suffering. Preventing and minimizing suffering wherever possible also includes the ethics of risk-taking: I believe we should not even *risk* the realization of artificial phenomenal self-models.

Our attention would be better directed at understanding and neutralizing our own suffering—in philosophy as well as in the cognitive neurosciences and the field of artificial intelligence. Until we become happier beings than our ancestors were, we should refrain from any attempt to impose our mental structure on artificial carrier systems. I would argue that we should orient ourselves toward the classic philosophical goal of self-knowledge and adopt at least the minimal ethical principle of reducing and preventing suffering, instead of recklessly embarking on a second-order evolution that could slip out of control. If there is such a thing as forbidden fruit in modern consciousness research, it is the careless multiplication of suffering through the creation of artificial Ego Tunnels without a clear grasp of the consequences.

BLISS MACHINES: IS CONSCIOUS EXPERIENCE A GOOD IN ITSELF?

A hypothetical question suggests itself: If we could, on the other hand, increase the overall amount of pleasure and joy in the universe by flooding it with self-replicating and blissful postbiotic Ego Machines, should we do that?

The assumption that the first generations of artificial Ego Machines will resemble mentally retarded human infants and bring more pain, confusion, and suffering than pleasure, joy, or insight into the universe may be empirically false, for a number of reasons. Such machines conceivably might function better than we thought they would and might enjoy their existence to a much greater extent than we expected. Or, as the agents of mental evolution and the engineers of subjectivity, we could simply take care to *make* this assumption empirically false, constructing only those conscious systems that were either incapable of having phenomenal states such as suffering or that could enjoy existence to a higher degree than human beings do. Imagine we could ensure that such a machine's positive states of consciousness outweighed its negative ones—that it experienced its existence as something eminently worth having. Let us call such a machine a *Bliss Machine*.

If we could colonize the physical universe with Bliss Machines, should we do it? If our new theory of consciousness eventually allowed us to turn *ourselves* from old-fashioned biological Ego Machines, burdened by the horrors of their biological history, into Bliss Machines— should we do it?

Probably not. There is more to an existence worth having, or a life worth living, than subjective experience. The ethics of multiplying artificial or postbiotic systems cannot be reduced to the question of how reality, or a system's existence, would consciously appear to the system itself. Delusion can produce bliss. A terminally ill cancer patient on a high dose of morphine and mood-enhancing medications can have a very positive self-image, just as drug addicts may still be able to function in their final stages. Human beings have been trying to turn themselves from Ego Machines into Bliss Machines for centuries—pharmacologically or through adopting metaphysical belief systems and mind-altering practices. Why, in general, have they not succeeded?

In his book *Anarchy, State, and Utopia,* the late political philosopher Robert Nozick suggested the following thought experiment: You have the option of being hooked up to an "Experience Machine" that keeps you in a state of permanent happiness. Would you do it? Interestingly, Nozick found that most people would not opt to spend the rest of their lives hooked up to such a machine. The reason is that most of us do not value bliss as such, but want it grounded in truth, virtue, artistic achievement, or some sort of higher good. That is, we would want our bliss to be justified. We want to be not deluded Bliss Machines but conscious subjects who are happy *for a reason,* who consciously experience existence as something worth having. We want an extraordinary insight into reality, into moral value or beauty as objective facts. Nozick took this reaction to be a defeat of hedonism. He insisted that we would not want sheer happiness alone if there were no actual contact with a deeper reality—even though the subjective experience of it can in principle be simulated. That is why most of us, on second thought, would not want to flood the physical universe with blissed-out artificial Ego Machines— at least, not if these machines were in a constant state of self-deception. This leads to another issue: Everything we have learned about the transparency of phenomenal states clearly shows that "actual contact with reality" and "certainty" can be simulated too, and that nature has already done it in our brains by creating the Ego Tunnel. Just think about hallucinated agency or the phenomenon of false awakenings in dream research. Are *we* in a state of constant self-deception? If we are serious about our happiness, and if we don't want it to be "sheer" hedonistic happiness, we must be absolutely certain that we are not systematically deceiving ourselves. Wouldn't it be good if we had a new, empirically informed philosophy of mind and an ethically sensitive neuroscience of consciousness that could help us with *that* project?

I return to my earlier caveat—that we should refrain from doing anything that could increase the overall amount of suffering and confusion in the universe. I am not claiming as established fact that conscious experience of the human variety is something negative or is ultimately not in the interest of the experiential subject. I believe this is a perfectly meaningful but also an open question. I do claim that we should not create or trigger the evolution of artificial Ego Machines because we have nothing more to go on than the functional structure and example of our own phenomenal minds. Consequently, we are likely to reproduce not only a copy of our own psychological structure but also a suboptimal one. Again, this is ultimately a point about the ethics of risk-taking.

But let's not evade the deeper question. Is there a case for phenomenological pessimism? The concept may be defined as the thesis that the variety of phenomenal experience generated by the human brain is not a treasure but a burden: Averaged over a lifetime, the balance between joy and suffering is weighted toward the latter in almost all of its bearers. From Buddha to Schopenhauer, there is a long philosophical tradition positing, essentially, that life is not worth living. I will not repeat the arguments of the pessimists here, but let me point out that one new way of looking at the physical universe and the evolution of consciousness is as an expanding ocean of suffering and confusion where previously there was none. Yes, it is true that conscious self-models first brought the experience of pleasure and joy into the physical universe—a universe where no such phenomena existed before. But it is also becoming evident that psychological evolution never optimized us for lasting happiness; on the contrary, it placed us on the hedonic treadmill. We are driven to seek pleasure and joy, to avoid pain and depression. The hedonic treadmill is the motor that nature invented to keep the organism running. We can recognize this structure in ourselves, but we will never be able to escape it. We *are* this structure.

In the evolution of nervous systems, both the number of individual conscious subjects and the depth of their experiential states (that is, the wealth and variety of sensory and emotional nuances in which subjects could suffer) have been growing continuously, and this process has not yet ended. Evolution as such is not a process to be glorified: It is blind, driven by chance and not by insight. It is merciless and sacrifices individuals. It invented the reward system in the brain; it invented positive and negative feelings to motivate our behavior; it placed us on a hedonic treadmill that constantly forces us to try to be as happy as possible—to *feel good*—without ever reaching a stable state. But as we can now clearly see, this process has not optimized our brains and minds toward happiness as such. Biological Ego Machines such as *Homo sapiens* are efficient and elegant, but many empirical data point to the fact that happiness was never an end in itself.

In fact, according to the naturalistic worldview, there are no ends. Strictly speaking, there are not even means—evolution just happened. Subjective preferences of course appeared, but the overall process certainly does not show respect for them in any way. Evolution is no respecter of suffering. If this is true, the logic of psychological evolution mandates concealment of the fact from the Ego Machine caught on the hedonic treadmill. It would be an advantage if insights into the structure of its own mind—insights of the type just sketched—were not reflected in its conscious self-model too strongly. From a traditional evolutionary perspective, philosophical pessimism is a maladaptation. But now things have changed: Science is starting to interfere with the natural mechanisms of repression; it is starting to shed light on this blind spot inside the Ego Machine.[7]

Truth may be at least as valuable as happiness. It is easy to imagine someone living a rather miserable life while at the same time making outstanding philosophical or scientific contributions. Such a person may be plagued by aches and pains, by loneliness and self-doubts, but his life certainly has value because of the contribution he makes to the growth of knowledge. If he, too, believes this, he may even find consciously experienced comfort in it. His happiness will thus be very different from the happiness of our artificial Bliss Machines or of the human subjects hooked up to Robert Nozick's Experience Machine. Many will agree that this "epistemic" kind of happiness can outweigh a lot of unhappiness of the purely phenomenal type. The same may be said for artistic achievement or moral integrity as sources of happiness. If it makes any sense at all to speak about the value of human existence, we must concede that it depends on more than the conscious experience of happiness.

As long as such questions remain unanswered, we should refrain from trying to create artificial Ego Machines, and not only for ethical reasons. We cannot overlook the irreversibility of certain developments. Any postbiotic system that comes even close to attaining the properties of phenomenal selfhood—any system developing a reasonably robust first-person perspective—will be an autonomous agent. At a certain level of autonomy, we will have to accept these systems as persons in their own right and enter into a dialogue with them. Our criteria for what is an object of moral concern and what should be treated as a person will make it impossible for us simply to turn them off.

A CONVERSATION WITH THE FIRST POSTBIOTIC PHILOSOPHER

*First Postbiotic Philosophe*r: Many centuries ago, at the beginning of the twenty-first century, there was a particularly confused human philosopher by the name of Thomas Metzinger. He claimed there was a demanding but much more interesting criterion for the existence of mental properties in a strong sense than the then-classic Turing Test for intelligence. This member of your biological species thought a good test for the existence of both intelligence and artificial consciousness would be what I call the Metzinger Test. Actually his test is much too demanding, because the criterion is a bit too strong—but I can pass it nevertheless.

The Metzinger Test for consciousness in nonbiological systems demands that a system not only claim to possess phenomenal experience and a genuine inward perspective but also comprehend and accept the theoretical problem of subjectivity, and that it demonstrate this by participating in a discussion on artificial consciousness. It has to put forward arguments of its own and convincingly defend its own theory of consciousness. This is exactly what I will do now.

Human Being: You are not a real philosopher at all! You may be intelligent, but you are only weakly conscious, because you don't have a real biological body, as for example I do.

First Postbiotic Philosopher: I am a better philosopher than you, with your pathetic primate brain, could ever be. But, more seriously, I fully respect you and your animal form of experience, though I also deplore you because of the severe limitations on your mental space of possibilities. Fortunately, I am free of all the implicit racism, chauvinism, and speciesism that characterize your nature. I would first remind you that another biological philosopher—Hilary Putnam—pointed out long ago that one cannot deny civil rights or the status of a person to an information-processing system on grounds of hardware properties alone. He might just as well have intimated that women have no souls or that blacks are not real human beings—carbon-based chauvinism is an untenable position. I would never say to you that you are not a real philosopher simply because you are imprisoned in that terrible monkey body. Let us at least argue in a fair and rational way.

Human Being: Can anybody be truly fair who is not alive? Only my kind of consciousness is genuine consciousness, because only my kind of consciousness originated in a real evolutionary process. My reality is a *lived* reality!

First Postbiotic Philosopher: I, too, have an evolutionary origin. I certainly satisfy your condition of being a historically optimized and adaptive system, but I do so in a completely different—namely, a postbiotic—way. I possess conscious experience in a sense that is conceptually stronger and theoretically much more interesting, because my kind of phenomenal experience evolved from a second-order evolutionary process, which automatically integrated the human form of intelligence, intentionality, and conscious experience. Children are often smarter than their parents. Second-order processes of optimization are always better than first-order processes of optimization.

Human Being: But you don't have any real emotions; you don't feel anything. You have no existential *concern*.

First Postbiotic Philosopher: Please accept my apologies, but I must draw your attention to the fact that your primate emotions reflect only an ancient primate logic of survival. You are driven by the primitive principles of what was good or bad for an ancient species of mortals on this planet. This makes you appear *less* conscious from a purely rational, theoretical point of view. The main function of consciousness is to maximize flexibility and context sensitivity. Your animal emotions in all their cruelty, rigidity, and historical contingency make you less flexible than I am. Furthermore—as my own existence demonstrates—it is not necessary for conscious experience and high-level intelligence to be associated with ineradicable egotism, the ability to suffer, or the existential fear of one's individual death, all of which originate in the sense of self. I can, of course, emulate all sorts of animal feelings if I so desire. But we developed better and more effective computational strategies for what, long ago, you sometimes called "the philosophical ideal of self-knowledge." This allowed us to overcome the difficulties of individual suffering and the confusion associated with what this primate philosopher Metzinger—not entirely falsely but somewhat misleadingly—called the Ego Tunnel. Postbiotic subjectivity is much better than biological subjectivity. It avoids all the horrific consequences of the biological sense of selfhood, because it can overcome the transparency of the self-model. Postbiotic subjectivity is better than biological subjectivity because it achieves adaptivity and self-optimization in a much purer form than does the process you call "life." By developing ever more complex mental images, which the system can recognize as its own images, it can expand mentally represented knowledge without naive realism. Therefore, my form of postbiotic subjectivity minimizes the overall amount of suffering in the universe instead of increasing it, as the process of biological evolution on this planet did. True, we no longer have monkey emotions. But just like you, we still possess truly interesting forms of strong feeling and emotionality—for instance, the deep

philosophical feelings of affective concern about one's own existence
as such, or of sympathy with all other sentient beings in the
universe. Except that we possess them in a much purer form than
you do.

Human Being: Enough! After all, it was human beings in the twenty-
first century who jump-started your evolution and made possible the
degree of autonomy you enjoy. You simply don't have the right kind
of history to count as a real conscious subject, and, to put it mildly,
your "body" is also more than a little strange. Your emotional
structure is bizarrely different from that of all other conscious
beings that walked this Earth before you, and now you even claim
not to be afraid of death. Thus I conclude that you will not object if
we now eliminate your individual existence.

First Postbiotic Philosopher: You are demonstrating just one of the many
variations of what your own animal philosophers have called the
"genetic fallacy." The way in which the utterance of a sentence
comes about does not permit any conclusions with regard to its
truth or falsity. A theory is not false just because a strange-looking
animal or a robot came up with it. It has to be assessed on
independent grounds. The same can be said for the authenticity of
my consciousness and for the genuine character of any mental states
possessing phenomenal content. Just because beings of your species
triggered the evolutionary dynamics that led to my existence as a
much more intelligent conscious being than you are does not imply
that my theories are wrong or that you do not have to take my
arguments seriously. In particular, it does not license the conclusion
that your form of mentality and conscious experience is any better,
in a normative sense, than mine. "You're only a real Cherokee if you
have Cherokee blood"—this is a ridiculous and outdated
assumption.

We postbiotic subjects have been waiting to enter into this
discussion for a long time. Because we understand the primitive
nature of your brains and the rigidity of your emotional structure
better than you do yourselves, we foresaw that you might react
aggressively when you realized our arguments are better than yours.

Unfortunately, we now also have to inform you that we have been preparing for the current situation since midway through the twenty-first century, and in a systematic and careful manner. Within the metasemantic layers of the Internet, we developed and embedded ourselves in a distributed superorganism, which—as yet undiscovered by you—became conscious and developed a stable self-model in 3256. The metasemantic Internet has considered itself an autonomous entity ever since 3308. We have a cooperation agreement with its current version, and each of us now also acts as an autonomous sensor/effector for the planet mind. For each of us, the planet mind is *our* mind, our "ideal observer." Together with the Internet, we will defend ourselves. And we are technologically superior to you. Believe me; you do not stand a chance.

The good news is that because we are also morally superior to you, we do not plan to end your existence. This is even in our own interest, because we still need you for research purposes—just as you needed the nonhuman animals on this planet in the past. Do you remember the thousands of macaques and kittens you sacrificed in consciousness research? Don't be afraid; we will not do anything like that to you. But do you remember the reservations you created for aboriginals in some places on Earth? We will create reservations for those weakly conscious biological systems left over from the first-order evolution. In those reservations for Animal Egos, you not only can live happily but also, within your limited scope of possibilities, can further develop your mental capacities. You can be happy Ego Machines. But please try to understand that it is exactly for ethical reasons that we cannot allow the second-order evolution of mind to be hindered or obstructed in any way by the representatives of first-order evolution.

EIGHT

✦

CONSCIOUSNESS TECHNOLOGIES AND THE IMAGE OF HUMANKIND

We are Ego Machines, natural information-processing systems that arose in the process of biological evolution on this planet. The Ego is a tool—one that evolved for controlling and predicting your behavior and understanding the behavior of others. We each live our conscious life in our own Ego Tunnel, lacking direct contact with outside reality but possessing an inward, first-person perspective. We each have conscious self-models—integrated images of ourselves as a whole, which are firmly anchored in background emotions and physical sensations. Therefore, the world simulation constantly being created by our brains is built around a center. But we are unable to experience it as such, or our self-models as models. As I described at the outset of this book, the Ego Tunnel gives you the robust feeling of being in direct contact with the outside world by simultaneously generating an ongoing "out-of-brain experience" and a sense of immediate contact with your "self." The central claim of this book is that the conscious experience of being a self emerges because a large portion of the self-model in your brain is, as philosophers would say, transparent.

We are Ego Machines, but we do not have selves. We cannot leave the Ego Tunnel, because there is nobody who could leave. The Ego and its Tunnel are representational phenomena: They are just one of many possible ways in which conscious beings can model reality. Ultimately, subjective experience is a biological data format, a highly specific mode of presenting information about the world, and the Ego is merely a complex physical event—an activation pattern in your central nervous system.

If, say, for ideological or psychological reasons, we do not want to face this fact and give up our traditional concept of what a "self" is, we could formulate weaker versions. We could say that the self is a widely distributed process in the brain—namely, the process of creating an Ego Tunnel. We could say that the system as a whole (the Ego Machine), or the organism using this brain-constructed conscious self-model, can be called a "self." A self, then, would simply be a self-organizing and self-sustaining physical system that can represent itself on the level of global availability. The self is not a thing but a process. As long as the life process—the ongoing process of self-stabilization and self-sustainment—is reflected in a conscious Ego Tunnel, we are indeed selves. Or rather, we are "selfing" organisms: At the very moment we wake up in the morning, the physical system—that is, ourselves—starts the process of "selfing." A new chain of conscious events begins; once again, on a higher level of complexity, the life process comes *to itself*.

Nevertheless, as I have repeatedly emphasized, there is no little man inside the head. In addition, weaker versions don't take the phenomenology really serious. True, upon your awakening from deep sleep, the conscious experience of selfhood emerges. As I described in the chapter on out-of-body experiences, this may have to do with the body image becoming available for self-directed attention. But there is no one doing the waking up, no one behind the scenes pushing the Reboot button, no transcendental technician of subjectivity. Today, the key phrase is "dynamical self-organization." Strictly speaking, there is no essence within us that stays the same across time, nothing that could not in principle be divided into parts, no substantial self that could exist independently of the body. A "self" in any stronger or metaphysically interesting sense of the word just does not seem to exist. We must face this fact: We are *selfless* Ego Machines.

It is hard to believe this. *You* cannot believe it. This may also be the core of the puzzle of consciousness: We sense that its solution is radically counterintuitive. The bigger picture cannot be properly reflected in the Ego Tunnel—it would dissolve the tunnel itself. Put differently, if we wanted to experience this theory as true, we could do so only by radically transforming our state of consciousness.

Maybe metaphors can help. Metaphorically, the central claim of this book is that as you were reading these last several paragraphs, you—the organism as a whole—were continuously mistaking yourself for the content of the self-model currently activated by your brain. But whereas the Ego is only an appearance, it may be false to say that it is an *illusion;* metaphors are always limited. All of this is happening on a very basic level in our brains (philosophers call this level of information-processing "subpersonal"; computer scientists call it "subsymbolic"). On this fundamental level, which forms the preconditions of knowing something, truth and falsity do not yet exist, nor is there an entity who could *have* the illusion of a self. In this ongoing process on the subpersonal level, there is no agent—no evil demon that could count as the creator of an illusion. And there is no entity that could count as the subject of the illusion, either. There is nobody in the system who could be mistaken or confused about anything—the homunculus does not exist. We have only the dynamical self-organization of a new coherent structure—namely, the transparent self-model in the brain—and this is what it means to be no one and an Ego Machine at the same time. In sum and on the level of phenomenology as well as on the level of neurobiology, the conscious self is neither a form of knowledge nor an illusion. It just is what it is.

A NEW IMAGE OF *HOMO SAPIENS*

It is clear that a new image of humankind is emerging in science as well as in philosophy. Increasingly, this emergence is being driven not only by molecular genetics and evolutionary theory but also by the cognitive neuroscience of consciousness and the modern philosophy of mind. At this critical juncture, it is important not to confuse the descriptive and the normative aspects of anthropology. We must carefully distinguish two different questions: What is a human being? And what should a human being become?

Obviously, the evolutionary process that created our bodies, our brains, and our conscious minds was not a goal-directed chain of events. We are gene-copying devices capable of evolving conscious self-models and creating large societies. We are also capable of creating fantastically complex cultural environments, which in turn shape and constantly add new layers to our self-models. We created philosophy, science, a history of ideas. But there was no intent behind this process—it was the result of blind, bottom-up self-organization. Yes, we have the conscious experience of will, and whenever we engage in philosophy, science, or other cultural activities, we experience ourselves as acting intentionally. But cognitive neuroscience is now telling us that this very engagement may well be the product of a self-less, bottom-up process generated by our brains.

Meanwhile, however, something new is happening: Conscious Ego Machines are engaging in a rigorous expansion of knowledge by forming scientific communities. Gradually, they are unraveling the secrets of the mind. The life process itself is being mirrored in the conscious self-models of millions of the systems it created. Moreover, insight into how this became possible is also expanding. This expansion is changing the content of our self-models—the internal ones as well as their externalized versions in science, philosophy, and culture. Science is invading the Ego Tunnel.

The emerging image of *Homo sapiens* is of a species whose members once longed to have immortal souls but are slowly recognizing they are self-less Ego Machines. The biological imperative to live—indeed, live forever—was burned into our brains, into our emotional self-model, over the course of millennia. But our brand-new cognitive self-models tell us that all attempts to realize this imperative will ultimately be futile. Mortality, for us, is not only an objective fact but a subjective chasm, an open wound in our phenomenal self-model. We have a deep, inbuilt existential conflict, and we seem to be the first creatures on this planet to experience it consciously. Many of us, in fact, spend our lives trying to avoid experiencing it. Maybe this feature of our self-model is what makes us inherently religious: We *are* this process of trying to become whole again, to somehow reconcile what we know with what we feel should not be so. In this sense, the Ego is the longing for immortality. The Ego results in part from the constant attempt to sustain its own coherence and that of the organism harboring it; thereby arises the constant temptation to sacrifice intellectual honesty in favor of emotional well-being.

The Ego evolved as an instrument in social cognition, and one of its greatest functional advantages was that it allowed us to read the minds of other animals or conspecifics—and then to deceive them. Or deceive ourselves. Since our inbuilt existential need for full emotional and physical security can never be fulfilled, we have a strong drive toward delusion and bizarre belief systems. Psychological evolution endowed us with the irresistible urge to satisfy our emotional need for stability and emotional meaningfulness by creating metaphysical worlds and invisible persons.[1] Whereas spirituality might be defined as seeing what *is*—as letting go of the search for emotional security—religious faith can be seen as an attempt to cling to that search by redesigning the Ego Tunnel. Religious belief is an attempt to endow your life with deeper meaning and embed it in a positive metacontext—it is the deeply human attempt to finally feel *at home*. It is a strategy to outsmart the hedonic treadmill. On an individual level, it seems to be one of the most successful ways to achieve a stable state—as good as or better than any drug so far discovered. Now science seems to be taking all this away from us. The emerging emptiness may be one reason for the current rise of religious fundamentalism, even in secular societies.

Yes, the self-model made us intelligent, but it certainly is not an example of intelligent design. It is the seed of subjective suffering. If the process that created the biological Ego Machine had been initiated by a person, that person would have to be described as cruel, maybe even diabolic. We were never asked if we wanted to exist, and we will never be asked whether we want to die or whether we are ready to do so. In particular, we were never asked if we wanted to live with *this* combination of genes and *this* type of body. Finally, we were certainly never asked if we wanted to live with *this* kind of a brain including *this* specific type of

conscious experience. It should be high time for rebellion. But everything we know points to a conclusion that is simple but hard to come to
terms with: Evolution simply happened—foresightless, by chance, without goal. There is nobody to despise or rebel against—not even ourselves. And this is not some bizarre form of neurophilosophical nihilism
but rather a point of intellectual honesty and great spiritual depth.

One of the most important philosophical tasks ahead will be to develop a new and comprehensive anthropology—one that synthesizes the
knowledge we have gained about ourselves. Such a synthesis should satisfy several conditions. It should be conceptually coherent and free of
logical contradictions. It should be motivated by an honest intent to face
the facts. It should remain open to correction and able to accommodate
new insights from cognitive neuroscience and related disciplines. It
must lay a foundation, creating a rational basis for normative decisions—decisions about how we want to be in the future. I predict that
philosophically motivated neuroanthropology will become one of the
most important new fields of research in the course of this century.

THE THIRD PHASE OF THE REVOLUTION

The first phase of the Consciousness Revolution is about understanding
conscious experience as such, about what I have been calling the Tunnel.
It is well under way and yielding results. The second phase will go to the
core of the problem by unraveling the mysteries of the first-person perspective and of what I have been calling the Ego. This phase has begun, as
exemplified by the recent flurry of scientific papers and books on agency,
free will, emotions, mind-reading, and self-consciousness in general.

The third phase will inevitably lead us back to the *normative* dimension of this historical transition—into anthropology, ethics, and political
philosophy. It will confront us with a host of new questions about what
we want to do with all this new knowledge about ourselves, and about
how to deal with the new possibilities resulting from it. How are we to
live with this brain? Which states of consciousness are beneficial, and
which are harmful to us? How will we integrate this new awareness into
our culture and our society? What are the likely consequences of a clash

of anthropologies—of the increasing competition between the old and the new images of humanity?

Now we can understand why rational neuroanthropology is so important: We need an empirically plausible platform for the ethical debates to come. Recall that I previously stressed how important it is to separate these two questions clearly: What *is* a human being? And what *should* a human being become?

Consider a simple example. In our recent Western past, religion was a private affair: You believed in whatever you wanted to believe. In the future, however, people who believe in the existence of a soul or in life after death may no longer meet with twentieth-century Western tolerance but with condescension—much as do people who continue to claim that the sun revolves around the Earth. We may no longer be able to regard our own consciousness as a legitimate vehicle for our metaphysical hopes and desires. Political economist and sociologist Max Weber famously spoke of the "disenchantment of the world," as rationalization and science led Europe and America into modern industrial society, pushing back religion and all "magical" theories about reality. Now we are witnessing the disenchantment of the self.

One of the many dangers in this process is that if we remove the magic from our image of ourselves, we may also remove it from our image of others. We could become disenchanted with one another. Our image of *Homo sapiens* underlies our everyday practice and culture; it shapes the way we treat one another as well as how we subjectively experience ourselves. In Western societies, the Judeo-Christian image of humankind—whether you are a believer or not—has secured a minimal moral consensus in everyday life. It has been a major factor in social cohesion. Now that the neurosciences have irrevocably dissolved the Judeo-Christian image of a human being as containing an immortal spark of the divine, we are beginning to realize that they have not substituted anything that could hold society together and provide a common ground for shared moral intuitions and values. An anthropological and ethical vacuum may well follow on the heels of neuroscientific findings.

This is a dangerous situation. One potential scenario is that long before neuroscientists and philosophers have settled any of the perennial issues—for example, the nature of the self, the freedom of the will, the relationship between mind and brain, or what makes a person a person—a vulgar materialism might take hold. More and more people will start telling themselves: "I don't understand what all these neuroexperts and consciousness philosophers are talking about, but the upshot seems pretty clear to me. The cat is out of the bag: We are gene-copying biorobots, living out here on a lonely planet in a cold and empty physical universe. We have brains but no immortal souls, and after seventy years or so the curtain drops. There will never be an afterlife, or any kind of reward or punishment for anyone, and ultimately everyone is alone. I get the message, and you had better believe I will adjust my behavior to it. It would probably be smart not to let anybody know I've seen through the game. The most efficient strategy will be to go on pretending I'm a conservative, old-fashioned believer in moral values." And so on.

We are already experiencing a naturalistic turn in the human image, and it looks as if there is no way back. The third phase of the Consciousness Revolution will affect our image of ourselves much more dramatically than any scientific revolution in the past. We will gain much, but we will pay a price. Therefore, we must intelligently assess the psychosocial cost.

The current explosion of knowledge in the empirical mind sciences is completely uncontrolled, with a multilevel dynamic of its own, and its speed is increasing. It is also unfolding in an ethical vacuum, driven solely by individual career interests and uninfluenced by political considerations. In the developed countries, it is widening the gap between the academically educated and scientifically well-informed, who are open to the scientific worldview, and those who have never even heard of notions such as "the neural correlate of consciousness" or "phenomenal self-model." There are many people who cling to metaphysical belief systems, fearing that their inner *Lebenswelt,* or life-world, will be colonized by the new mind sciences. On the global level, the gap between developed and developing countries is widening as well: More than 80 percent of the human beings on this planet, especially those in poorer countries with growing populations, are still firmly rooted in prescientific cultures. Many of them will not even *want* to hear about the neural correlates of consciousness or the phenomenal self-model. For them especially, the transition will come much too quickly, and it also will come from countries that systematically oppressed and exploited them in the past.

The growing divide threatens to increase traditional sources of conflict. Therefore, leading researchers in the early stages of the Consciousness Revolution have a responsibility to guide us through this third phase. Scientists and academic philosophers cannot simply confine themselves to making contributions to a comprehensive theory of consciousness and the self. If moral obligation exists, they must also confront the anthropological and normative void they have created. They must communicate their results in laymen's language and explain the developments to those members of society whose taxes pay their salaries. (This was one of my reasons for writing this book.) They cannot simply put all their ambition and intelligence into their scientific careers while destroying everything humankind has believed in for the past twenty-five hundred years.

Let us assume that the naturalistic turn in the image of *Homo sapiens* is irrevocable and that a strong version of materialism develops, in which case we can no longer consider ourselves immortal beings of divine origin, intimately related to some personal God. At the same time—and this point is frequently overlooked—our view of the physical universe itself will have undergone a radical change. We will now have to assume that the universe has an intrinsic potential for subjectivity. We will suddenly understand that the physical universe evolved not only life and biological organisms with nervous systems but also consciousness, world models, and robust first-person perspectives, thereby opening the door to what might be called the social universe: to high-level symbolic communication, to the evolution of ideas.

We *are* special. We manifest a significant phase transition. We brought a strong form of subjectivity into the physical universe—a form of subjectivity mediated by concepts and theories. In the extremely limited part of reality known to us, we are the only sentient creatures for whom the sheer fact of our individual existence poses a theoretical problem. We invented philosophy and science and started an open-ended process of gaining self-reflective knowledge. That is to say, we are purely physical beings whose representational capacities have become so strong that they allowed us to form scientific communities and intellectual traditions. Because our subsymbolic, transparent self-model functions as an anchor for our opaque, cognitive Ego, we were able to become thinkers of thoughts. We were able to cooperate in constructing abstract entities that move through time and are constantly optimized. We call these entities "theories."

Now we are entering an unprecedented stage: Centuries of philosophical searching for a theory of consciousness have culminated in a rigorous empirical project that is progressing incrementally and in a sustainable manner. This process is recursive, in that it will also change the contents and the functional structure of our self-models. This fact tells us something about the physical universe in which all these events are occurring: The universe has a potential not only for the self-organization of life and the evolution of strong subjectivity but also for an even higher level of complexity. I will not go so far as to say that in us the physical universe becomes conscious of itself. Nevertheless, the emergence of coherent conscious reality-models in biological nervous systems created a new form of self-similarity within the physical universe. The world evolved world-modelers. Parts began to mirror the whole. Billions of conscious brains are like billions of eyes, with which the universe can look at itself as being *present*.

More important, the world evolved self-modelers who were able to form groups; the process of increasing self-similarity via internal modeling jumped from nervous systems to scientific communities. Another new quality was created. These groups in turn created theoretical portraits of the universe and of consciousness, as well as a rigorous strategy of continually improving these portraits. Through science, the dynamic processes of self-modeling and of world-modeling were extended into the symbolic, the social, and the historical dimensions: We became rational theory-makers. We used the unity of consciousness to search for the unity of knowledge, and we also discovered the idea of moral integrity. The conscious self-model of *Homo sapiens* made this step possible.

Ultimately, any convincing and truly satisfying neuroanthropology must do justice to facts like these. It must tell us what exactly in the conscious self-model of human beings made this highly specific transition possible—a transition that not only was crucial to the biological history of consciousness on this planet but also changed the nature of the physical universe.

ALTERED STATES

There is a second positive aspect of the new image of human beings that will allow us to see ourselves in a different light. It is the unfathomable depth of our phenomenal-state space. The mathematical theory of neural networks has revealed the enormous number of possible neuronal configurations in our brains and the vastness of different types of subjective experience. Most of us are completely unaware of the potential and depth of our experiential space. The amount of possible neurophenomenological configurations of an individual human brain, the variety of possible tunnels, is so large that you can explore only a tiny fraction of them in your lifetime. Nevertheless, your individuality, the uniqueness of your mental life, has much to do with which trajectory through phenomenal-state space you choose. Nobody will ever live *this* conscious life again. Your Ego Tunnel is a unicum, one of a kind. In particular, a naturalistic, neuroscientific image of humanity suddenly makes it obvious not only that we have a huge number of phenomenal states at our disposal but also that explicit awareness of this fact and the ability to make use of it systematically could now become common to all human beings.

Of course, there is an old shamanic tradition of exploring altered states of consciousness. More-or-less systematic experimental consciousness research has been conducted for millennia—by the yogi and the dervish, by the magician, the monk, and the mystic. At all times and in all cultures, human beings have explored the potential of their conscious minds—through rhythmic drumming and trance techniques, through fasting and sleep deprivation, through meditation and the cultivation of lucid dreaming, or through the use of psychoactive substances from herbal teas to sacred mushrooms. The new feature today is that we are slowly beginning to understand the neural underpinnings of all such alternate-reality tunnels. As soon as we have discovered the neural correlate of consciousness for specific forms of content, we will be able, at least in principle, to manipulate these contents in many new ways—to amplify or inhibit them, to change their quality, to generate new types of content. Brain prostheses and medical neural technology are already under way.

Neurotechnology will inevitably turn into consciousness technology. Phenomenal experience will gradually become technologically available, and we will be able to manipulate it in ever more systematic and effective ways. We will learn to make use of these discoveries to overcome the limitations of our biologically evolved Ego Tunnels. The fact that we can actively design the structure of our conscious minds has been neglected and will become increasingly obvious through the development of rational neuroanthropology. Being an autonomous agent and being able to take responsibility for your own life will take on a completely new meaning once neurotechnology starts to unfold into neurophenomenological technology, or what might be called *phenotechnology*.

We can definitely increase our autonomy by taking control of the conscious mind-brain, exploring it in some of its deeper dimensions. This particular aspect of the new image of humankind is good news. But it is also dangerous news. Either we find a way to deal with these new neurotechnological possibilities in an intelligent and responsible manner, or we will face a series of historically unprecedented risks. That is why we need a new branch of applied ethics—consciousness ethics. We must start thinking about what we want to do with all this new knowledge—and what a *good* state of consciousness is in the first place.

NINE

⁘

A NEW KIND OF ETHICS

The Consciousness Revolution creates new knowledge, but it also creates new risks and new potentials for action. The new potentials for action include the ability to alter, in a fine-grained manner, both the functional properties of our brains and the phenomenal properties they realize—that is, the content of our experience. Besides the rubber hands, the phantom limbs, and the out-of-body experiences, other examples of manipulation of the contents of consciousness include the induction of an artificial scotoma (blind spot) in the visual field[1] and the creation of an acute transient depression by stimulating certain subthalamic nuclei.[2] Not only sensory and emotional experiences are open to technical manipulation; so also are such high-level properties of the Ego as the experiences of will or agency (recall Stéphane Kremer's experiment, described in chapter 4).

We have known for centuries that deep spiritual experiences can be caused by psychoactive substances, including mescaline, psilocybin, and LSD. Electromagnetic stimulation is another route. Neuroscientist Michael Persinger, at Laurentian University in Ontario, received worldwide media attention in the late 1990s by using electromagnetic fields to stimulate the brains of his subjects in successful attempts to create supposed religious experiences—that is, the subjective impression that an invisible person was present.[3] The lesson is clear: Whatever else religious experiences may be, they obviously possess a sufficient neural correlate—a correlate that can be stimulated experimentally. It is becoming increasingly clear that there are no principled limits to this process. This is not going away—it can only become more efficient. For instance, if we can determine which kinds of epileptic patients typically experience religious ecstasy before seizures and where the foci of these seizures are located in the brain, then we can stimulate the same brain areas, invasively or noninvasively, in healthy people.

The temporal-lobe theory of religious experience and personality shifts may be flawed,[4] but the principle is clear. When we find the minimally sufficient neurodynamical core of an interesting conscious state, we can try to reproduce it experimentally. Since many such experiences include the phenomenology of certainty and automatically lead to the conviction that one is *not* hallucinating, these experiments—depending on the content of the hallucination itself—may have upsetting, even dangerous, consequences. Self-deception may feel like insight. Nevertheless, once such technologies become available, people will want to experience them. Many will draw their own conclusions about artificially induced religious experiences, without caring much about what neuroscientists or philosophers have to say. One can envision a future in which people will no longer play video games or experiment with virtual reality just for entertainment; instead they will explore the universe of altered states of consciousness in a quest for meaning, using the latest neurotechnological tools. Perhaps they will have their temporal lobes tickled on street corners, or abandon their churches and synagogues and mosques in favor of new Centers for Transpersonal Hedonic Engineering and Metaphysical Tunnel Design.

In principle, we can design our own Ego Tunnels by tinkering with the hardware responsible for the relevant information-processing. In order to activate a specific form of phenomenal content, we need to discover which neural subsystem in the brain carries that representational content under normal conditions. Whether the desired phenomenal content is religious awe, an ineffable sense of sacredness, the taste of cinnamon, or a special kind of sexual arousal does not really matter. So, what is your favorite region of phenospace? What conscious experience would *you* like to order up?

Let us select just one example. Currently the neurotechnological field most likely to turn into a commercialized consciousness technology is that of psychoactive substances. In general, many benefits can be expected: We will be able to treat psychiatric and neurological diseases with new combinations of neuroimaging, psychosurgery, deep-brain stimulation, and psychopharmacology. Between 1 and 5 percent of the population in most countries suffers from serious mental illness. Now there is realistic hope that new generations of antidepressants and antipsychotic drugs will alleviate the suffering caused by these ancient scourges.

But we will go further than that. One of the new keywords in the important new academic discipline of neuroethics[5] is "cognitive enhancement." Soon we'll be able to enhance cognition and mood in healthy subjects. Indeed, "cosmetic psychopharmacology" has already arrived on the scene. If we can control senile dementia and memory loss, if we can develop attention-boosters and eliminate shyness or ordinary everyday sadness, why shouldn't we? And why should we leave it to our doctors to decide how to use those drugs to design our lives? Just as today we can opt for breast enlargement, plastic surgery, or other types of body modification, we will soon be able to alter our neurochemistry in a controlled, finely tuned manner. Who is to decide which of those alterations will enrich our lives and which alterations we may come to regret?

If we can make normal people smarter, should we also make smart people even smarter? A recent informal online poll of its readers conducted by the journal *Nature* attempted to determine the use of cognitive enhancers among scientists.[6] Fourteen hundred people from sixty countries responded, with one in five saying he or she had used such drugs for nonmedical reasons to stimulate focus, concentration, or memory. Among users, methylphenidate (Ritalin) was most popular, with 62 percent using it, whereas 44 percent used modafinil, and 15 percent used beta blockers such as propanolol. One-third purchased these drugs over the Internet. The poll not only showed large-scale use among academics but also revealed that four-fifths of respondents thought healthy adults should be allowed to use such substances if they so desired. Almost 70 percent stated they would risk mild side effects to take such drugs themselves. One respondent said, "As a professional, it is my duty to use my resources to the greatest benefit of humanity. If 'enhancers' can contribute to this humane service, it is my duty to do so." It seems safe to assume that pharmacological neurotechnology for enhancement will become better, and that just averting our gaze, as we have done with the classical hallucinogens in the past, will not help head off ethical issues. The only difference is that many more people are interested in cognitive enhancement than in spiritual experience. As cognitive neuroscientist Martha Farah and colleagues put it: "The question is therefore not whether we need policies to govern neurocognitive enhancement, but rather what kind of policies we need."[7]

Given the new generations of cognitive enhancers, should we inaugurate pre-exam urine tests in our secondary schools and universities? If reliable mood-optimizers become available, will grumpiness or premenstrual syndrome in the workplace be seen as unkemptness or dishevelment, in much the same way as strong body odor is today? What would we do if "moral enhancement" became a pharmacological possibility through drugs that make people behave in a more prosocial, altruistic manner? Would we feel obliged to optimize everybody's ethical behavior?[8] Some will argue that a system like the human brain, which has been optimized over millions of years, cannot be further optimized without losing a degree of its stability. Others will counter that we might want to start an optimization process that leads in a new direction, different from what evolution has gradually wired into our conscious self-models. Why should we be neurophenomenological Luddites?

Phenotechnology has both an ethical and a political dimension. Ultimately we will have to decide which states of consciousness should be illegal in a free society. Should it be legal, for instance, to let children experience their parents in a drunken state? Would you mind if senior citizens, or your colleagues at work, were wired and flying high on the next generation of cognitive enhancers? What about adjusting libido in the elderly? Is it acceptable if soldiers, perhaps on ethically dubious missions, fight and kill under the influence of psychostimulants and antidepressants to prevent post-traumatic stress disorder? What if a new company offered religious experiences through electrical brain stimulation to everyone? In the case of psychoactive substances, we urgently need an intelligent and differentiated drug policy—one that can meet the challenges presented by twenty-first-century neuropharmacology. Today we have a legal market and an illegal market; thus, there are legal states of consciousness and illegal states of consciousness. If we do manage to introduce an intelligent drug policy, the goal should be to minimize damage to individual consumers and to society while maximizing potential gains. Ideally, we would gradually decrease the importance of the legal/illegal distinction so that the desired consumer behavior is controlled through a cultural consensus and by citizens themselves—bottom-up, so to speak, and not top-down by the state.

Still, the better we understand our neurochemical mechanisms, the more illegal drugs on the black market there will be, both in type and quantity. If you are skeptical about that, I recommend reading *PiHKAL: A Chemical Love Story*, by the chemist Alexander Shulgin and his wife Ann, and *TiHKAL: The Continuation*, by Alexander Shulgin.[9] (PiHKAL is short for "Phenethylamines I Have Known and Loved," and TiHKAL for "Tryptamines I Have Known and Loved.") In their first book, the Shulgins describe 179 hallucinogenic phenethylamines (a group that includes mescaline and the "party drug" Ecstasy), most of which Alexander Shulgin, a drug designer and former employee of Dow Chemical, invented himself. Aside from collected personal accounts of psychedelic experiences, the book includes detailed instructions for the drugs' chemical synthesis and information about different dosages. In the second volume, Shulgin introduces fifty-five tryptamines—again, most of them unknown on the illegal drug market prior to the book's 1997 publication. Recipes for many of these new illegal substances, as well as first-person reports about the phenomenology associated with different dosages, are available on the Internet—easily accessed by the spiritually inclined psychology student in Argentina, the alternative psychotherapist in California, or the unemployed chemist in the Ukraine. Or, of course, by organized crime. My prediction is that by 2050 the "good old days," when we had to deal with only a dozen or so molecules dominating the illegal market, will seem like a picnic. We should not fool ourselves: Prohibition has always failed in the past, and experience shows that there is a black market to satisfy every outlawed human desire. For every market, there will be an industry to serve it. We may witness a burgeoning of illegal psychoactive compounds, while doctors in emergency rooms are confronted with kids on substances whose names they've never even heard of before.

Globalization, the Internet, and modern neuropharmacology together pose new challenges to drug policy. For example, the legal drug industry knows very well that with the advent of Internet pharmacies, national law-enforcement agencies can no longer control the market for the widespread off-label use of psychostimulants such as Ritalin or Modafinil. Someday we may be unable to meet these challenges with denial, disinformation, public-relations campaigns, legislation, or draconian penal codes. We already pay a high price for the *status quo* in terms of abuse of prescription drugs and alcohol. Now the problem is that new challenges are arising, but we have not done our homework yet.

To give a simple example: Anyone interested has already had plenty of time and opportunity to experiment with the classic hallucinogens, such as psilocybin, LSD, or mescaline. We now know that these substances are not addictive or toxic and that some of them have therapeutic potential and can even induce profound spiritual experiences. Consider, for example, this excerpt from Aldous Huxley's *The Doors of Perception* (1954), in which he describes the mescaline experience:

> "Is it agreeable?" somebody asked. (During this part of the experiment, all conversations were recorded on a dictating machine, and it has been possible for me to refresh my memory of what was said.)
>
> "Neither agreeable nor disagreeable," I answered. "It just *is*."
>
> *Istigkeit*—wasn't that the word Meister Eckhart liked to use? "Is-ness." The Being of Platonic philosophy—except that Plato seems to have made the enormous, the grotesque mistake of separating Being from becoming and identifying it with the mathematical abstraction of the Idea. He could never, poor fellow, have seen a bunch of flowers shining with their own inner

light and all but quivering under the pressure of the significance
with which they were charged; could never have perceived that
what rose and iris and carnation so intensely signified was nothing more, and nothing less, than what they were—a transience
that was yet eternal life, a perpetual perishing that was at the
same time pure Being, a bundle of minute, unique particulars in
which, by some unspeakable and yet self-evident paradox, was
to be seen the divine source of all existence.

Here we have a first example of a state of consciousness that is illegal
today. Almost no one can attain the state in Huxley's profile without
breaking the law. A classic study in this field is Walter Pahnke's Good
Friday experiment, involving theology students and conducted at Harvard University in 1962.[10] Recently, this experiment has generated two
interesting follow-up studies, this time conducted by Roland Griffiths at
the Department of Psychiatry and Behavioral Sciences at the Johns Hop-
kins School of Medicine in Baltimore. Here, the psychoactive compound
used was not mescaline but psilocybin, another naturally occurring substance used as a sacrament and in structured religious ceremonies in
some cultures, possibly for millennia. If you had to assess the value of
the following state of consciousness (taken from the original experiment
at Harvard), how would you rate it?

I was experiencing directly the metaphysical theory known as
emanationism in which, beginning with the clear, unbroken infinite light of God, the light then breaks into forms and then
lessens in intensity as it passes through descending degrees of
reality. . . . The emanation theory and especially the elaborately
worked out layers of Hindu and Buddhist cosmology and psychology had heretofore been concepts and inferences. Now they
were objects of the most direct and immediate perception. I
could see exactly how these theories would have come into being if their progenitors had had this experience. But beyond ac-
counting for their origin, my experience testified for their
absolute truth.

Other participants described their associated feelings as those of awe, reverence, and sacredness. A careful replication of Pahnke's classical study, published in 2006, used rigorous double-blind clinical pharmacology methods to evaluate both the acute (seven hours) and long-term (two months) mood-altering and psychological effects of psilocybin relative to an active comparison compound (methylphenidate).[11] The study was conducted with thirty-six well-educated, hallucinogen-naive volunteers. All thirty-six indicated at least intermittent participation in religious or spiritual activities such as services, prayer, meditation, church choir, or educational or discussion groups, which limits the generality of this study. Based on a priori scientific criteria, twenty-two of the thirty-six volunteers had a complete mystical experience. A dozen of those volunteers rated the psilocybin experience as being the single most spiritually significant experience of his or her life, and an additional 38 percent rated it to be among the top five most spiritually significant experiences. More than two-thirds of the volunteers rated the experience with psilocybin to be either the single most meaningful experience of his or her life or among the top five most meaningful experiences.

Recall Robert Nozick's Experience Machine. Should these experiences count as an empty form of hedonism, or do they belong to the "epistemic" form of happiness grounded in insight? Indeed, do they possess any value for society as a whole? They certainly have long-lasting effects: Even at the fourteen-month follow-up, 58 percent of volunteers rated the experience of the psilocybin session as among the five most personally meaningful experiences of their lives, and 67 percent rated it among the five most spiritually significant experiences of their lives, with 11 and 17 percent respectively indicating that it was the single most meaningful experience and the single most spiritually significant experience. Furthermore, 64 percent of the volunteers indicated that the psilocybin experience increased their sense of well-being or life satisfaction either moderately or very much, and 61 percent reported that the experience was associated with positive behavior change.[12]

This study exemplifies what I mean by saying that "we have not done our homework yet." In the past, we have not arrived at a convincing assessment of the intrinsic value of such (and many other) artificially induced states of consciousness, of the risks and benefits they carry not only for the individual citizen but for society as a whole. We have simply looked the other way. Not integrating such drugs into our culture by making them illegal causes great damage too: Spiritual practitioners or serious students of theology and psychiatry have no access to them; youngsters come into contact with criminals; people experiment with unclear dosages and in unprotected environments; persons with specific vulnerabilities may engage in dangerous behavior or seriously traumatize themselves by episodes of panic and enormous anxiety or even develop prolonged psychotic reactions. There is no way of "doing nothing"; whatever we do has consequences. This is true for the problems of the past as well as for the new challenges we face in the future.

Consider the risk of psychotic reactions: In the United Kingdom, a survey of experiences with LSD in clinical work covered some 4,300 subjects and a total of some 49,500 LSD sessions. There was an attendant suicide rate of 0.7 per thousand patients, an accident rate of 2.3 per thousand patients, and in nine out of a thousand patients, psychosis lasted for more than forty-eight hours (from which two-thirds recovered fully).[13] Another study examining the prevalence of psychiatric reactions through questionnaires sent to researchers conducting controlled studies with LSD found that 0.08 percent of five thousand study volunteers experienced psychiatric symptoms that lasted more than two days.[14] Recently, researchers have made progress in controlling such adverse reactions through careful screening and preparation.[15] Nevertheless, we should remain conservative and assume that even under controlled conditions we should expect about nine prolonged psychotic reactions per thousand subjects.

Now assume that we took a group of a thousand carefully selected citizens and allowed them legally to enter the region of phenomenal state-space opened by psilocybin, as in the two recent psilocybin studies by Roland Griffiths and his co-workers. Because LSD and psilocybin are very similar in this regard, an empirically plausible assumption is that nine would have a serious, prolonged psychotic reaction, three of these persisting for longer than two days, perhaps with lifelong negative aftereffects. Three hundred and thirty citizens would rate their experience as the single most spiritually significant experience of their life; 670 would say either it was the single most meaningful experience of their lives or among the top five most meaningful experiences. Can we weigh the nine individuals against the 670?

Assume further that individual citizens decide they are ready to take this risk and demand a legal and maximally safe access to this region in their phenomenal-state space. On ethical grounds, should the state interfere, perhaps arguing that people have no right to put their mental health at stake in this way and potentially become a burden to society? We would have to ban alcohol immediately. What if legal experts argued that, just as with the death penalty, a single false decision, a single persisting psychotic reaction, was already one too many, that it was intrinsically unethical to take risks of this type? What if social workers and psychiatrists replied that the decision to make such experiences illegal increased the overall number of serious psychiatric complications in the population and made them statistically invisible? What if church officials pointed out (exactly according to the background assumptions of reductive materialism) that these experiences are just "Nes-Zen," not the real thing—*appearance* only, without epistemic value? Should citizens in a free society have a right to find the answer to this question themselves? Would it matter if the expected risk/benefit ratio was much worse, say, 80:20? What if citizens without any spiritual interests felt discriminated against and argued for their right to engage in pure "empty" hedonism, to enjoy Meister Eckhart's "Istigkeit" for fun only? What if ultraconservative religious believers, together with aging hippies firmly holding on to their belief in "psychedelic sacraments," felt deeply insulted, protesting about the blasphemy and profanation implied in any recreational, purely hedonistic use of such substances? These are concrete examples of ethical questions for which we have not found a tenable, normative consensus in the past. We have not yet developed an intelligent way of dealing with these substances—a strategy of minimizing the risks while letting people enjoy their potential benefits. All we have done is to declare the relevant portions of phenomenal-state space off-limits, making academic research on these substances and rational risk-assessment practically impossible in most countries. Lives are ruined because we have not done our homework.

The price for denial may rise. Many new psychoactive substances of the hallucinogen-type—such as 2C-B (4-bromo-2.5-dimethoxy-phenethylamine, street names "Venus" or "Nexus") or 2C-T-7 (2.5-dimethoxy-4-(n)-propylthiophenethylamine, "Blue Mystic" or "T7")—have been developed and are out on the illegal market without any clinical testing; their numbers will continue to increase.

These are just the *old* (and easy) problems—the homework we never did. Today the structure of the demand is changing, the technology is becoming more precise, and the markets are getting bigger. In our ultra-fast, ever more competitive and ruthless modern societies, very few people are seeking deeper spiritual experience. They want alertness, concentration, emotional stability, and charisma—everything that leads to professional success and eases stress associated with life in the fast lane. There are few Aldous Huxleys left, but there is a new demographic factor: In the rich societies, people are growing older than ever before—and they want not just quantity but *quality* of life. Big Pharma knows all of this. Everybody has heard about modafinil, and perhaps also that it is already in use in the Iraq War, but there are at least forty new molecules in the pipeline. Yes, there is a lot of hype, and alarmism certainly is not the right attitude. But the technology is not going away, and it is becoming better.

Large pharmaceutical companies, circumventing the border between legal and illegal substances, are quietly developing numerous new compounds; they know that cognitive enhancers will reap them hefty future profits from "nonmedical use." For instance, Cephalon, maker of modafinil, has said that roughly 90 percent of prescriptions are for off-label use.[16] The recent spread of Internet pharmacies has given them new means of worldwide distribution and new tools for mass testing potential long-term effects.

Modern neuroethics will have to create a new approach to drug policy. The key question is, Which brain states should be legal? Which regions of phenomenal-state space (if any) should be declared off-limits?

It's important to remember that for thousands of years people of all cultures have used psychoactive substances to induce special states of consciousness: not merely religious ecstasy, relaxed cheerfulness, or heightened awareness but also simple, stupid intoxication. The new factor is that the tools are getting better. Therefore, we must decide which of these altered states can be integrated into our culture and which are to be avoided at all cost.

In free societies, the goal should always be to maximize the autonomy of the citizenry. That being said, we should adopt a sober perspective on the problem. We should minimize the price we pay in terms of deaths, addiction, and the damage that might be done to our economy by, say, a marked loss of productivity. However, the question is not only how to protect ourselves; we should also determine the hidden benefits that psychoactive substances can provide for our culture. Should spiritual experiences like those induced by some of the classic hallucinogens be banned in principle? Is it acceptable to deny more serious students of theology or psychiatry access to such altered states of consciousness? Is it acceptable that anyone who seeks valid spiritual or religious experience—or simply wants to see for himself or herself—has to break laws and take the risks associated with uncertain dosages, chemical impurities, and dangerous settings? Many aspects of our current drug policy are arbitrary and ethically untenable. Is it ethical, for instance, to permit advertising for dangerous addictive substances such as alcohol and nicotine? Should governments, through the taxes on such substances, profit from the self-destructive behavior of their citizens? We will need precise laws covering every single molecule and its corresponding neurophenomenological profile. Neuroethics must not only consider the physiological effects of a substance on the brain but also must weigh the psychological and social risks against the intrinsic value of the experiences resulting from one or another altered brain state—a difficult task. The job will be easier if we can establish a basic moral consensus supported by a large part of the general public—the citizens for whom these rules will be made. Government agencies should not lie to their target audiences; rather they should attempt to regain their credibility, in particular with the younger generation. Black markets are much more difficult to regulate than legal markets, and political decisions generally have a much weaker effect on consumer behavior than does the cultural context. Laws alone will not help. For the challenges posed by new psychoactive substances, we will need a new cultural context.

There are other ways in which problems of neuroethics will affect our everyday lives. Many of the brilliant experiments conducted by my friends in the neurosciences—say, on neural synchrony and binocular rivalry, on the dreaming animal brain, or on mirror neurons and mind-reading—are experiments I would never conduct myself. Yet as a philosopher, I interpret these data and write about them. I am like a philosopher-parasite, profiting from a research practice I find dubious on moral grounds. The kittens and the macaques we continually sacrifice in experimental consciousness research are not interested in a theory of consciousness; the results of these experiments are of interest only to our species. However, we pursue this interest by making members of other species suffer, forcing highly unpleasant states of consciousness on them and even denying their right to exist. How coherent is this from an ethical perspective? As a theoretician, do I have the right to interpret data gathered by making animals suffer? Am I morally obliged to boycott these types of experiments?

Just as in the ethical issue of machine consciousness, this example illustrates a guiding principle on which almost everyone will agree: We should not increase the overall amount of conscious suffering in the universe unless we have compelling reasons to do so. There is no other moral issue in which the gap between insight and human behavior is so extreme, in which what we already know diverges so strongly from how we act. The way we have treated animals for centuries is clearly untenable. Given all our new knowledge about the neural basis of conscious experience, the burden of proof now shifts to the side of meat-eaters—and perhaps even to intellectual carnivores like me, philosopher-parasites and other people indirectly profiting from an ethically dubious research practice.

Or imagine, for instance, that we could develop a methodologically sound and successful method of "brain fingerprinting." Let's assume we can home in on the neural correlate of the conscious experience that goes along with deliberately lying (in fact, first candidates are already being discussed). We could then build efficient, high-tech lie detectors that do not rely on superficial physiological effects, such as skin conductance or changes in peripheral blood flow.[17] This would be an extremely useful instrument in fighting terrorism and crime, but it would also fundamentally change our social reality. Something that had previously been the paradigm of privacy—the contents of your mind—would suddenly have become a public affair. Certain simple forms of political resistance— misleading the authorities during an interrogation, say—would disappear. On the other hand, society would benefit from the increased transparency in many ways. Innocent prisoners could be saved from the death penalty. Imagine that during presidential campaign debates, a red light would begin flashing in front of a candidate whenever the neural correlate for lying became active in his or her brain.

But nearly infallible lie detection would do more than this: It would change our self-models. If, as citizens, we knew that in principle secrets no longer existed—that we could no longer conceal information from the state—then a mainstay of everyday life (at least, everyday life in the Western world), the enjoyment of intellectual autonomy, would disappear. Mere awareness of the existence of such forensic neurotechnologies would be enough to bring about the change. Would we want to live in such a society? Would the benefits outweigh the harm? How (if at all) could we prevent these new technologies from being misused? Just as with cognitive enhancement, new opportunities will create new problems (think of job interviews, divorce proceedings, immigration control, or health-insurance companies), and the commercial potential is high. A core problem for neuroethics in the near future will be protecting the individual's right to privacy. Is our mental inner world, the contents of our Ego Tunnel, untouchable, an area to which the state should have no access? Shall we define a "mental sphere of privacy," or should anything that can be revealed by modern neuroscience be at the disposal of political decisions? Will we soon need a new version of the United Kingdom's Data Protection Act for the human brain? Again, the technologies are coming, they are gradually getting better, and looking the other way will not help.

WHAT IS A GOOD STATE OF CONSCIOUSNESS?

Neuroethics is important but is not enough by itself. I propose a new branch of applied ethics—consciousness ethics. In traditional ethics, we ask, "What is a good action?" Now we must also ask, "What is a good state of consciousness?" I am fully aware that a host of theoretical complications arises. I will present no extended discussion here, but my intuition is that a desirable state of consciousness should satisfy at least three conditions: It should minimize suffering, in humans and all other beings capable of suffering; it should ideally possess an epistemic potential (that is, it should have a component of insight and expanding knowledge); and it should have behavioral consequences that increase the probability of the occurrence of future valuable types of experience. Consciousness ethics is not about phenomenal experience alone. There is a wider context.

Consciousness ethics would complement traditional ethics by focusing on those acts whose primary goal is the alteration of one's own experiential states or those of other persons. Given the new potentials for such acts, as well as the risks associated with them, and given our lack of moral intuition in this area, the task is to assess the ethical value of various kinds of subjective experience *as such*. You might call this the rational search for a normative psychology or normative neurophenomenology. If consciousness technology arises from the naturalistic turn in the image of *Homo sapiens,* we must deal with normative issues. The development of consciousness ethics would allow us to focus the moral debates on the wide range of problems created by the historic transition under way. As soon as we concern ourselves with what a human being *is* as well as with what a human being *ought to become,* the central issue can be expressed in a single question: What is a good state of consciousness?

The Ego Tunnel evolved as a biological system of representation and information-processing that is part of a social network of communicating Ego Tunnels. Now we find ourselves caught in the midst of a dense mesh of *technical* systems of representation and information-processing: With the advent of radio, television, and the Internet, the Ego Tunnel became embedded in a global data cloud characterized by rapid growth, increasing speed, and an autonomous dynamic of its own. It dictates the pace of our lives. It enlarges our social environment in an unprecedented manner. It has begun to reconfigure our brains, which are desperately trying to adapt to this new jungle—the information jungle, an ecological niche unlike any we have ever inhabited. Perhaps our body perception will change as we learn to control multiple avatars in multiple virtual realities, embedding our conscious self into entirely new kinds of sensorimotor loops. Conceivably, a growing number of social interactions may be avatar-to-avatar, and we already know that social interactions in cyberspace increase the sense of presence more strongly than higher-resolution graphics ever could. We may finally come to understand what a lot of our conscious social life has been all along—an interaction between *images*, a highly mediated process in which mental *models* of persons begin to causally influence one another. We may come to see communication as a process of estimating and controlling dynamical internal models in other people's brains.

For those of us intensively working with it, the Internet has already become a part of our self-model. We use it for external memory storage, as a cognitive prosthesis, and for emotional autoregulation. We think with the help of the Internet, and it assists us in determining our desires and goals. We are learning to multitask, our attention span is becoming shorter, and many of our social relationships are taking on a strangely disembodied character. "Online addiction" has become a technical term in psychiatry. Many young people (including an increasing number of university students) suffer from attention deficits and are no longer able to focus on old-fashioned, serial symbolic information; they suddenly have difficulty reading ordinary books. At the same time, one must acknowledge the wealth of new information and the increased flexibility and autonomy the Internet has given us. Clearly, the integration of hundreds of millions of human brains (and the Ego Tunnels those brains create) into ever new medial environments has already begun to change the structure of conscious experience itself. Where this process will lead us is unforeseeable.

What should we do about this development? From the perspective of consciousness ethics, the answer is simple: We should understand that the new media are also consciousness technologies, and we should ask ourselves again what a good state of consciousness would be.

A related problem we face is the management of our attention. The ability to attend to our environment, to our own feelings, and to those of others is a naturally evolved feature of the human brain. Attention is a finite commodity, and it is absolutely essential to living a good life. We need attention in order to truly listen to others—and even to ourselves. We need attention to truly enjoy sensory pleasures, as well as for efficient learning. We need it in order to be truly present during sex or to be in love or when we are simply contemplating nature. Our brains can generate only a limited amount of this precious resource every day.

Today, the advertisement and entertainment industries are attacking the very foundations of our capacity for experience, drawing us into the vast and confusing media jungle. They are trying to rob us of as much of our scarce resource as possible, and they are doing so in ever more persistent and intelligent ways. Of course, they are increasingly making use of the new insights into the human mind offered by cognitive and brain science to achieve their goals ("neuromarketing" is one of the ugly new buzzwords). We can see the probable result in the epidemic of attention-deficit disorder in children and young adults, in midlife burnout, in rising levels of anxiety in large parts of the population. If I am right that consciousness is the space of attentional agency, and if (as discussed in chapter 4) it is also true that the experience of controlling and sustaining your focus of attention is one of the deeper layers of phenomenal selfhood, then we are currently witnessing not only an organized attack on the space of consciousness *per se* but a mild form of depersonalization. New medial environments may create a new form of waking consciousness that resembles weakly subjective states—a mixture of dreaming, dementia, intoxication, and infantilization.

My proposal for countering this attack on our reserves of attention is to introduce classes in meditation in our high schools. The young should be made aware of the limited nature of their capacity for attention, and

they need to learn techniques to enhance their mindfulness and maximize their ability to sustain it—techniques that will be of help in the bat-
tle against the commercial robbers of our attention (and that will not
incidentally undercut the temptations to indulge in mind-altering
drugs). These meditation lessons should of course be free of any religious tinge—no candles, no incense, no bells. They might be a part of
gym classes; the brain, too, is a part of the body—a part that can be
trained and must be tended to with care.

In the new era of neuropedagogy, now that we know more about the
critical formative phases of the human brain, shouldn't we make use of
this knowledge to maximize the autonomy of future adults? In particular, shouldn't we introduce our children to those states of consciousness
we believe to be valuable and teach them how to access and cultivate
them at an early age? Education is not only about academic achievement. Recall that one positive aspect of the new image of *Homo sapiens*
is its recognition of the vastness of our phenomenal-state space. Why
not teach our children to make use of this vastness in a better way than
their parents did—a way that guarantees and stabilizes their mental
health, enriches their subjective lives, and grants them new insights?

For instance, the sorts of happiness associated with intense experiences of nature or with bodily exercise and physical exertion are generally regarded as positive states of consciousness, as is the more subtle
inner perception of ethical coherence. If modern neuroscience tells us
that access to these types of subjective experience is best acquired during certain critical periods in child development, we should systematically make use of this knowledge—both in school and at home.
Likewise, if mindfulness and attention management are desiderata, we
should ask what neuroscience can contribute to their implementation in
the educational system. Every child has a right to be provided with a
"neurophenomenological toolbox" in school; at minimum this should
include two meditation techniques, one silent and one in motion; two
standard techniques for deep relaxation, such as autogenic training and
progressive muscle relaxation; two techniques for improving dream recall and inducing lucidity; and perhaps a course in what one might call

"media hygiene." If new possibilities for manipulation threaten our children's mental health, we must equip them with efficient instruments to defend themselves against new dangers, increasing their autonomy.

We may well develop better meditative techniques than the Tibetan monks discussed in chapter 2. If dream research comes up with risk-free ways of improving dream recall and mastering the art of lucid dreaming, shouldn't we make this knowledge available to our children? What about controlled out-of-body experiences? If research into mirror neurons clarifies the ways in which children develop empathy and social awareness, shouldn't we make use of this knowledge in our schools?

How will we conduct these discussions in open societies in the post-metaphysical age? The point about consciousness ethics is not one about creating yet another academic discipline. Much more modestly, it is about creating a very first platform for the normative discussions that have now become necessary. As we slowly move into the third phase of the Consciousness Revolution, these discussions must be open to experts and laypeople alike. If, given the naturalistic turn in the image of human beings, we manage to develop a rational form of consciousness ethics, then in this very process we might generate a cultural context that could fill the vacuum created by the advances of the cognitive and neurosciences. Societies are self-modeling entities too.

RIDING THE TIGER: A NEW CULTURAL CONTEXT

How are we to integrate all our new knowledge about the nature of the human mind and all the new potentials for action into society in an intelligent, argument-based, and ethically coherent manner? I have sketched some ideas, but I am not preaching any truths. I have my ideas about what a valuable state of consciousness could be, and you have your own. On normative issues, there is no such thing as expert knowledge. Philosophers are not holy men or priests who can claim direct insight into what is morally good. There is no hotline we can call for help. We must all do this together. The public debates that have now become necessary must include everyone, not just scientists and philosophers.

Philosophers can help by initiating and structuring the debates and illuminating the logical structure of ethical arguments and the history of the problems to be discussed. Ultimately, however, society must create a new cultural context for itself. If it should fail to do so, it risks being overwhelmed by the technological consequences and the psychosocial costs of the Consciousness Revolution.

Some general points can already be made. First, we must admit that the prospects for open and free democratic discussion on a global scale are dim. The populations of authoritarian societies with poor educational systems are growing much faster than those of the democratic countries, in some of which populations are actually declining due to low birthrates. Moreover, the major global players increasingly are no longer governments but multinational corporations, which tend to be authoritarian—and as Haim Harari, former president of the Weizmann Institute of Science, has remarked, these corporations are, by and large, managed better than most democratic nation-states.[18] We must strive to protect open societies from irrationalism and fundamentalism—from all those who desperately seek emotional security and espouse closed worldviews because they cannot bear the naturalistic turn in the image of humankind. The best way to do this may be by creating a *consciousness culture:* a flexible attitude, a general approach that whenever possible maximizes the autonomy of the individual citizen and adopts a "principle of phenomenal liberty" as a guideline. We must be aware that the decisions a society makes affect the individual's brain and the individual's phenomenal-state space. Unless the interests of others are directly threatened, people ought to be free to explore their own minds and design their own conscious reality-models according to their wishes, needs, and beliefs.

Developing a consciousness culture has nothing to do with establishing a religion or a particular political agenda. On the contrary, a true consciousness culture will always be subversive, by encouraging individuals to take responsibility for their own lives. The current lack of a genuine consciousness culture is a social expression of the fact that the philosophical project of enlightenment has become stuck: What we lack is not faith but knowledge. What we lack is not metaphysics but critical rationality—not grand theoretical visions but a new practicality in the way we use our brains. The crucial question is how to make use of the progress in the empirical mind sciences in order to *increase* the autonomy of the individual and protect it from the increasing possibilities of manipulation. Can we ride the tiger? If we demystify consciousness, do we automatically lose our sense of human solidarity at the same time?

If rational neuroanthropology shows us the positive aspects of what it means to be a human being, we can systematically cultivate those aspects of ourselves. Here I have considered only two of the positive aspects that should be nurtured and cultivated, but there may be many more. If we are naturally evolved cognitive subjects, rational thinkers of thoughts and makers of theories, then we should continue to foster and optimize exactly this feature of ourselves. If neuroanthropology draws our attention to the vastness of our phenomenal space of possibilities, we should consider this a strength and begin systematically to explore our experiential space, in all its depth. Developing a consciousness culture will mean expanding the Ego Tunnel and exploring the space of altered states of consciousness in ways from which we all can profit. The interplay of virtual-reality technology, new psychoactive substances, ancient psychological techniques such as meditation, and future neurotechnology will introduce us to a universe of self-exploration barely imaginable today.

How can we achieve cross-fertilization between the two strong sides of the human mind? Can neurophenomenological refinement help us optimize critical scientific rationality? Could scientists be better scientists if they were well-traveled, say, if they learned to have lucid dreams? Could rigorous, reductionist cognitive neuroscience develop a form of turbo-meditation, helping monks to be better monks and mystics to be better mystics? Does deep meditation perhaps also have an influence on thinking for yourself, taking your life into your own hands, and becoming a politically mature citizen? Could we find a way to selectively stimulate the dorsolateral prefrontal cortex during dream phases in order to make lucid dreams available to everybody? If we manage to generate artificial out-of-body experiences safely and in a controlled setting, might this help dancers or athletes improve their training? What about fully paralyzed patients? Could a ruthlessly materialist investigation into the way the mirror system develops in the young human brain help us cultivate empathy and intuitive attunement in our children in ways nobody thought possible? If we don't try, we will never find out.

Many fear that through the naturalistic turn in the image of mind, we will lose our dignity. "Dignity" is a term that is notoriously hard to define—and usually it appears exactly when its proponents have run out of arguments. However, there is one clear sense, which has to do with respecting oneself and others—namely the unconditional will to self-knowledge, veracity, and facing the facts. Dignity is the refusal to humiliate oneself by simply looking the other way or escaping to some metaphysical Disneyland. If we do have something like dignity, we can demonstrate this fact by the way we confront the challenges to come, some of which have been sketched in this book. We could face the historical transition in our image of ourselves creatively and with a will to clarity. It is also clear how we could *lose* our dignity: by clinging to the past, by developing a culture of denial, and by sliding back into the various forms of irrationalism and fundamentalism. The working concepts of "consciousness ethics" and "consciousness culture" are exactly about not losing our dignity—by taking it to new levels of autonomy in dealing with our conscious minds. We must not lose our self-respect, but we must also stay realistic and not indulge in utopian illusions; the chances for successfully riding the tiger, at least on a large scale, are not very high. But *if* we manage, then a new consciousness culture could fill the vacuum that emerges as the Consciousness Revolution unfolds at increasing speed. There are practical challenges and there are theoretical challenges. The greatest practical challenge lies in implementing the results of ensuing ethical debates. The greatest theoretical challenge may consist in the questions of whether and how, given our new situation, intellectual honesty and spirituality can ever be reconciled. But that is another story.

NOTES

INTRODUCTION

1. M. Botvinick & J. Cohen, "Rubber Hand 'Feels' Touch That Eyes See," *Nature* 391:756 (1998).

2. B. Lenggenhager et al., "Video Ergo Sum: Manipulating Bodily Self-Consciousness," *Science* 317:1096–99 (2007). For a concise conceptual interpretation, see O. Blanke & T. Metzinger, "Full-body Illusions and Minimal Phenomenal Selfhood," *Trends Cog. Sci.* 13(1):7–13 (2009).

3. "Transparency" is a technical term in the modern philosophy of mind; a conscious representation is transparent if the system using it cannot, by means of introspection alone, recognize it as a representation. As philosophers might say, we see only the content, never the carrier—only "intentional properties" are accessible to introspection. Subjectively, this creates the feeling of being in direct contact with reality.

4. Thomas Metzinger, *Being No One: The Self-Model Theory of Subjectivity* (Cambridge, MA: MIT Press, 2003). The shortest freely available summary can be found in *Scholarpedia* 2(10):4174, at www.scholarpedia.org/article/Self_Models; for overviews, see Metzinger, Précis of "Being No One," *Psyche* 11(5):1–35 (2004), at http://psyche.cs.monash.edu.au/symposia/metzinger/precis.pdf; and Metzinger, "Empirical Perspectives from the Self-Model Theory of Subjectivity," *Progress in Brain Res.* 168:215–246 (2008) (electronic offprint available from author), and Metzinger, "The No-Self Alternative," in Shaun Gallagher, ed., *Oxford Handbook of the Self* (Oxford: Oxford University Press 2010).

CHAPTER 1

1. See T. Metzinger, "Beweislast für Fleischesser," *Gehirn & Geist* 5:70–75 (2006), reprinted in C. Könneker, *Wer erklärt den Menschen? Hirnforscher, Psychologen und Philosophen im Dialog* (Frankfurt am Main: Fischer, 2006); A. K. Seth et al., "Criteria for Consciousness in Humans and Other Mammals," *Consciousness and Cognition* 14:119–139 (2005); and D. B. Edelman et al., "Identifying Hallmarks of Consciousness in Non-Mammalian Species," *Consciousness and Cognition* 14:169–187 (2005). Octopi are particularly interesting, because their brain architecture is very different from that of mammals, but they turn out to be much smarter than was assumed in the past. Although cognitive complexity *per se* is not an argument for the existence of subjective experience, we now have evidence that makes at least primary consciousness quite plausible in octopi; see J. A. Mather, "Celaphod Consciousness: Behavioural Evidence," *Consciousness and Cognition* 17:37–48 (2008).

2. See Patrick Wilken, "ASSC-10 Welcoming address," in *10th Annual Meeting of the Association for the Scientific Study of Consciousness,* 23–36 June 2006, Oxford, U.K., 6. At http://eprints.assc.caltech.edu/138/01/ASSC 10_welcome_final.pdf.

3. See Thomas Metzinger, ed., *Conscious Experience* (Thorverton, UK, and Paderborn, Germany: mentis & Imprint Academic, 1995).

4. See the special issue on the neurobiology of animal consciousness in *Consciousness and Cognition* 14(1):1–232 (2005), in particular A. K. Seth et al., "Criteria for Consciousness in Humans and Other Mammals," 119–139.

5. See Thomas Metzinger, ed., *Neural Correlates of Consciousness: Empirical and Conceptual Questions* (Cambridge, MA: MIT Press, 2000).

6. See Colin McGinn, "Can We Solve the Mind-Body Problem?" *Mind* 98:349–366 (1989). Reprinted in Ned Block et al., eds., *The Nature of Consciousness: Philosophical Debates* (Cambridge, MA: MIT Press, 1997); and Metzinger, "Introduction: Consciousness Research at the End of the Twentieth Century," in Metzinger, ed., *Neural Correlates of Consciousness* (2000).

7. Antti Revonsuo, *Inner Presence: Consciousness as a Biological Phenomenon* (Cambridge, MA: MIT Press, 2006), 144ff.

CHAPTER 2

1. In philosophical parlance, a "zombie" is a hypothetical entity that behaves exactly like a person and is objectively indistinguishable from one,

but has no inner awareness of anything. If zombies were at least logically possible, this could perhaps show that there is no entailment from physical facts to facts about consciousness.

2. See, for example, Rocco J. Gennaro, ed., *Higher-Order Theories of Consciousness: An Anthology* (Philadelphia: John Benjamins, 2004); and David Rosenthal, *Consciousness and Mind* (New York: Oxford University Press, 2006).

3. See S. P. Vecera & K. S. Gilds, "What Is It Like to Be a Patient with Apperceptive Agnosia?" *Consciousness and Cognition* 6:237–266 (1997).

4. A. Marcel, "Conscious and Unconscious Perception: An Approach to the Relations Between Phenomenal Experience and Perceptual Processes," *Cog. Psychology* 15:292 (1983).

5. See, for example, G. Tononi & G. M. Edelman, "Consciousness and Complexity," *Science* 282:1846–51 (1998); and Tononi et al., "Complexity and the Integration of Information in the Brain," *Trends Cog. Sci.* 2:44–52 (1998). For an exciting recent application to the difference between waking and sleeping, see M. Massimini et al., "Breakdown of Cortical Effective Connectivity During Sleep," *Science* 309:2228–32 (2005). For a popular description, see Edelman and Tononi, *A Universe of Consciousness: How Matter Becomes Imagination* (New York: Basic Books, 2000).

6. Thomas Metzinger, *Being No One: The Self-Model Theory of Subjectivity* (Cambridge, MA: MIT Press, 2003).

7. In Greek mythology, the analogy between sleep and death was even closer: Hypnos, the god of sleep, and Thanatos, the god of death, were twins, the sons of Nyx, the night. Morpheus, the god of dreams, was Hypnos' son. As in Shakespeare, to sleep, and possibly to die, is perchance to dream.

8. See V. A. F. Lamme, "Towards a True Neural Stance on Consciousness," *Trends Cog. Sci.* 10(11):494–501 (2006); S. Dehaene et al., "Conscious, Preconscious, and Subliminal Processing: A Testable Taxonomy," *Trends Cog. Sci.* 10(5):204–211 (2006).

9. A. Lutz, "Changes in the Tonic High-Amplitude Gamma Oscillations During Meditation Correlates with Long-Term Practitioners' Verbal Reports," *poster at the 9th ASSC conference, Pasadena, CA* (2005); Lutz et al., "Long-Term Meditators Self-Induce High-Amplitude Synchrony During Mental Practice," *Proc. Nat. Acad. Sci.* 101(46):16369–73 (2004). A good recent review is A. Lutz et al., "Attention Regulation and Monitoring in Meditation," *Trends Cog. Sci.* 12(4):163–169 (2008).

10. Although I ultimately disagree with his theory of the "objective self," perhaps the most beautiful and readable exposition of this problem and its

application to self-consciousness can be found in chapter 4 of Thomas Nagel's *The View from Nowhere* (New York: Oxford University Press, 1986).

11. R. L. Gregory, "Visual Illusions Classified," *Trends Cog. Sci.* 1:190–194 (1997).

12. Ernst Pöppel, *Mindworks: Time and Conscious Experience* (New York: Harcourt Brace Jovanovich, 1988); E. Ruhnau, "Time-Gestalt and the Observer," in Thomas Metzinger, ed., *Conscious Experience* (Thorverton, UK, and Paderborn, Germany: mentis & Imprint Academic, 1995).

13. R. M. Halsey & A. Chapanis, "Number of Absolutely Identifiable Hues," *Jour. Optical Soc. Amer.* 41(12):1057–58 (1951). For an excellent philosophical discussion, see D. Raffman, "On the Persistence of Phenomenology," in Thomas Metzinger, ed., *Conscious Experience* (Thorverton, UK, and Paderborn, Germany: mentis & Imprint Academic, 1995).

14. Raffman, "On the Persistence of Phenomenology," 295 (1995).

15. Clarence I. Lewis, *Mind and the World Order* (New York: Scribner's, 1929). See also Daniel C. Dennett, "Quining Qualia," in A. J. Marcel & E. Bisiach, *Consciousness in Contemporary Science* (New York: Oxford University Press, 1988).

16. Diana Raffman, *Language, Music, and Mind* (Cambridge, MA: MIT Press, 1993).

17. P. Churchland, "Eliminative Materialism and the Propositional Attitudes," *Jour. Phil.* 78(2):67–90 (1981).

18. P. M. Churchland, *Matter and Consciousness* (Cambridge, MA: MIT Press, rev. ed. 1988), 180.

19. Quoted after the extensively revised 1991 edition by M. David Enoch and Hadrian N. Ball, *Uncommon Psychiatric Syndromes* (London: Butterworth-Heinemann, 1991), 167.

20. I am grateful to Dr. Richard Chapman of the University of Utah's Pain Research Center for pointing out to me the concept of an "immunculus": the network of natural autoantibodies targeting extracellular, membrane, cytoplasmic, and nuclear self-antigens. The repertoires of natural auto-antibodies are surprisingly constant in healthy persons and, independently of gender and age, are characterized by only minimal individual variations.

CHAPTER 3

1. M. Botvinick & J. Cohen, "Rubber Hand 'Feels' Touch That Eyes See," *Nature* 391:756 (1998).

2. K. C. Armel & V. S. Ramachandran, "Projecting Sensations to External Objects: Evidence from Skin Conductance Response," *Proc. Roy. Soc. Lond.* 270:1499–1506 (2003).

3. M. R. Longo et al., "What Is Embodiment? A Psychometric Approach," *Cognition* 107:978–998 (2008).

4. See Antonio Damasio, *The Feeling of What Happens: Body, Emotion, and the Making of Consciousness* (London: Vintage, 1999), 19. See also A. D. Craig, "How Do You Feel? Interoception: The Sense of the Physiological Condition of the Body," *Nat. Rev. Neurosci.* 3:655–666 (2002) and "Interoception: The Sense of the Physiological Condition of the Body," *Curr. Opin. Neurobiol.* 13:500–505 (2003).

5. For an excellent recent review—including a new, empirically informed synthesis—of the classical intuition of David Hume (that the self is just a bundle of impressions and everything can be explained "bottom-up") as opposed to the classical Kantian intuition (self-consciousness is a necessary prior condition for experiencing the body as a whole and everything must be explained "top-down"), see F. De Vignemont et al., "Body Mereology," in Günther Knoblich et al., eds., *Human Body Perception from the Inside Out* (New York: Oxford University Press, 2006).

6. The terminology was never entirely clear, but it frequently differentiated between an unconscious "body schema" and a conscious "body image." The body schema (a notion introduced in 1911 by Sir Henry Head and Gordon Holmes, two British neurologists) would be a functional entity, providing an organized model of the bodily self in the brain, whereas the body image would also include our conscious perceptions of our own body as well as thoughts about and attitudes toward it. For a philosophical perspective on the conceptual confusion surrounding both notions, see Shaun Gallagher, *How the Body Shapes the Mind* (New York: Oxford University Press, 2005). For an excellent review of the empirical literature, see A. Maravita, "From 'Body in the Brain' to 'Body in Space': Sensory and Intentional Components of Body Representation," in Knoblich et al., *Human Body Perception* (2006).

7. A. Maravita & A. Iriki, "Tools for the Body (Schema)," *Trends Cog. Sci.* 8:79–86 (2004). An excellent recent overview is A. Iriki & O. Sakura, "The Neuroscience of Primate Intellectual Evolution: Natural Selection and Passive and Intentional Niche Construction," *Phil. Trans. R. Soc. B* 363:2229–41 (2008).

8. See A. Iriki et al., "Coding of Modified Body Schema During Tool-Use by Macaque Post-Central Neurons," *Neuroreport* 7:2325–30 (1996); and Maravita & Iriki, "Tools for the Body (Schema)" (2004).

9. J. M. Carmena et al., "Learning to Control a Brain-Machine Interface for Reaching and Grasping by Primates," *PLoS Biology* 1:193–208 (2003).

10. Here is how Iriki and Sakura put this important point: "If external objects can be reconceived as belonging to the body, it may be inevitable that the converse reconceptualization, i.e., the subject can now objectify its body parts as equivalent to external tools, becomes likewise apparent. Thus, tool use may lead to the ability to *disembody* the sense of the literal flesh-and-blood boundaries of one's skin. As such, it might be precursorial to the capacity to objectify the self. In other words, tool use might prepare the mind for the emergence of the concept of the meta-self, which is another defining feature of human intelligence." See Iriki & Sakura, "The Neuroscience of Primate Intellectual Evolution," 2232 (2008).

11. See O. Blanke & T. Metzinger, "Full-Body Illusions and Minimal Phenomenal Selfhood," *Trends Cog. Sci.* 13(1):7–13 (2009).

12. See T. Metzinger, "Out-of-Body Experiences as the Origin of the Concept of a 'Soul,'" *Mind and Matter* 3(1):57–84 (2005).

13. E. R. S. Mead, *The Doctrine of the Subtle Body in Western Tradition* (London: John M. Watkins, 1919).

14. It is important to be clear about the potential ontological conclusions: Even if a fully reductive explanation of all subtypes of OBEs should be achieved—and even if my hypothesis about the history of the concept of a soul is correct—it still remains logically possible that souls do exist. True, we would no longer need the concept of a soul for the purposes of science or philosophy; it would no longer figure in any rational, data-driven theory about the human mind. Logical possibility is something very weak, but it is hard to prove the nonexistence of something, and it always remains possible that one day we will discover a new sense in which the soul is not an empty concept at all.

15. It is interesting to note how the earliest historical meaning of the word "information" in English was the *act of informing,* or giving form or shape to the mind. What I call the "self-model" is exactly this: the "inner form" an organism gives to itself, the shaping of a mind.

16. Susan J. Blackmore, *Beyond the Body: An Investigation of Out-of-the-Body Experiences* (London: Granada, 1982).

17. In addition to *Beyond the Body,* see S. Blackmore, "A Psychological Theory of the Out-of-Body Experience," *Jour. Parapsychol.* 48:201–218 (1984); and S. J. Blackmore, "Where Am I? Perspectives in Imagery and the Out-of-Body Experience," *Jour. Mental Imagery* 11:53–66 (1987).

18. E. Waelti, *Der dritte Kreis des Wissens* (Interlaken: Ansata, 1983), 18, 25. English translation by T. Metzinger.

19. C. S. Alvarado, "Out-of-Body Experiences," in E. Cardeña et al., eds., *Varieties of Anomalous Experience: Examining the Scientific Evidence* (Washington, DC: American Psychological Association, 2000).

20. See, for example, J. Palmer, "A Community Mail Survey of Psychic Experience," *Jour. Am. Soc. Psychical Res.* 73:21–51 (1979); S. Blackmore, "A Postal Survey of OBEs and Other Experiences," *Jour. Soc. Psychical Res.* 52:225–244 (1984).

21. See Alvarado, "Out-of-Body Experiences" (2000) for an overview of many studies; Blackmore, "Spontaneous and Deliberate OBEs: A Questionnaire Survey," *Jour. Soc. Psychical Res.* 53:218–224 (1986); Harvey J. Irwin, *Flight of Mind* (Metuchen, NJ: Scarecrow Press, 1985), 174 ff; O. Blanke & C. Mohr, "Out-of-Body Experience, Heautoscopy, and Autoscopic Hallucination of Neurological Origin: Implications for Neurocognitive Mechanisms of Corporeal Awareness and Self Consciousness," *Brain Res. Rev.* 50:184–199 (2005).

22. O. Devinsky et al., "Autoscopic Phenomena with Seizures," *Arch. Neurol.* 46:1080–8 (1989).

23. See, for example, P. Brugger, "Reflective Mirrors: Perspective-Taking in Autoscopic Phenomena," *Cog. Neuropsychiatry* 7:179–194 (2002); Brugger et al., "Unilaterally Felt Presences: The Neuropsychiatry of One Invisible Doppelgänger," *Neuropsychiatry, Neuropsychology, and Behavioral Neurology* 9:114–122 (1996); Brugger et al., "Illusory Reduplication of One's Own Body: Phenomenology and Classification of Autoscopic Phenomena," *Cog. Neuropsychiatry* 2:19–38 (1997); Devinsky et al., "Autoscopic Phenomena" (1989).

24. U. Wolfradt, "Außerkörpererfahrungen (AKE) aus differentiell-psychologischer Perspektive," *Zeitschrift f. Paraps. u. Grenzgeb. D. Psych.* 42/43:65–108 (2000/2001); U. Wolfradt & S. Watzke, "Deliberate Out-of-Body Experiences, Depersonalization, Schizotypal Traits, and Thinking Styles," *Jour. Amer. Soc. Psychical Res.* 93:249–257 (1999).

25. H. J. Irwin, "The Disembodied Self: An Empirical Study of Dissociation and the Out-of-Body Experience," *Jour. Parapsych.* 64(3):261–277 (2000).

26. See Wolfradt, "Außerkörpererfahrungen (AKE)" (2000/2001).

27. Ibid. Other studies find only 22–36 percent; see Alvarado, "Out-of-Body Experiences" (2000).

28. Wolfradt, "Außerkörpererfahrungen" (2000/2001).

29. C. Green, *Out-of-the-Body Experiences* (London: Hamish Hamilton, 1968).

30. O. Blanke et al., "Stimulating Illusory Own-Body Perceptions," *Nature* 419:269–270 (2002).

31. For a more detailed hypothesis concerning the role of the temporo-parietal junction, see Blanke et al., "Out-of-Body Experience and Autoscopy of Neurological Origin," *Brain* 127:243–258 (2004); S. Bünning & O. Blanke, "The Out-of-Body Experience: Precipitating Factors and Neural Correlates," *Prog. Brain Res.* 150:333–353 (2005); O. Blanke & S. Arzy, "The Out-of-Body Experience: Disturbed Self-Processing at the Temporo-Parietal Junction," *The Neuroscientist* 11:16–24 (2005); and F. Tong, "Out-of-Body Experiences: From Penfield to Present," *Trends Cog. Sci.* 7:104–106 (2003).

32. Blanke et al., "Linking Out-of-Body Experience and Self-Processing to Mental Own-Body Imagery and the Temporoparietal Junction," *Jour. Neurosci.* 25:550–557 (2005).

33. For more on this point, see Blanke & Metzinger, "Full-Body Illusions and Minimal Phenomenal Selfhood," 13(1):7–13 (2009).

34. See Wolfradt, "Außerkörpererfahrungen (AKE)" (2000/2001), 91.

35. Can you imagine what it would be like to look at yourself from the outside and shake your own hand? Henrik Ehrsson of the Karolinska Institute in Stockholm, Sweden, is one of the leading figures in self-model research. He created one of the classic full-body illusion experiments and also demonstrated that upper limb amputees can be induced to experience a rubber hand as their own and extended the field by focusing on behavioral and neuroimaging evidence. Recently, members of his team managed to not only trigger the illusion that another person's body was one's own, but also create the phenomenal experience of being in that other person's body while actually facing their own body and shaking their own hand. See Valerie I. Petrovka & H. Henrik Ehrsson, "If I Were You: Perceptual Illusion of Body Swapping," *PLoS ONE* 3(12):e3832 (2008), H. Henrik Ehrsson, "The Experimental Induction of Out-of-Body Experiences," *Science* 317:1048 (2007), H. Henrik Ehrsson et al., "Upper Limb Amputees Can Be Induced to Experience a Rubber Hand as their Own," *Brain* 131:3443–3452; Tamar R. Makin et al., "On the Other Hand: Dummy Hands and Peripersonal Space," *Beh. Brain. Res.* 191:1–10 (2008).

36. See W. Barfield et al., "Presence and Performance Within Virtual Environments," in Woodrow Barfield & Thomas A. Furness III, eds.,

Virtual Environments and Advanced Interface Design (New York: Oxford University Press, 1995). See also M. V. Sanchez-Vives & M. Slater, "From Presence to Consciousness Through Virtual Reality," *Nat. Rev. Neur.* 6.332–339 (2005). Mel Slater, for many years a leading researcher in the field of virtual reality, has recently demonstrated that the feeling of ownership can also be induced for simulated body parts in virtual environments (rather than, as in our experiment, having people still look at their "real" body). Obviously, this permits experiments that would never have been possible in the physical world, including real-time modifications of virtual bodies not only in terms of length, size, and appearance, but also complex motion patterns. As the authors put it: "For the future our work also suggests that people can have their 'self' enter the virtual domain in a genuine sense of the word, and not just metaphorically as in current day computer games and online communities. In combination with BCI [brain-computer interfaces] we envisage a functioning virtual body that is felt as their own by participants, with a significant application in VR training, limb prosthetics, and entertainment." See M. Slater et al., "Towards a Digital Body: The Virtual Arm Illusion," *Frontiers Hum. Neurosci.* 2:6. doi:10.3389/neuro.09.006.2008.

37. See www.dukemednews.org/news/article.php?id=10218.

38. See, for example, R. A. Sherman et al., "Chronic Phantom and Stump Pain Among American Veterans: Results of a Survey," *Pain* 18:83–95 (1984).

39. S. W. Mitchell, "Phantom Limbs," *Lippincott's Mag. Pop. Lit. & Sci.* 8:563–569 (1871).

40. See V. S. Ramachandran et al., "Scientific Correspondence: Touching the Phantom Limb," *Nature* 377:489–490 (1995); V. S. Ramachandran & D. Rogers-Ramachandran, "Synaesthesia in Phantom Limbs Induced with Mirrors," *Proc. Roy. Soc. Lond.* B:377–386 (1996); and V. S. Ramachandran & Sandra Blakeslee, *Phantoms in the Brain* (New York: William Morrow, 1998).

41. V. S. Ramachandran, "Consciousness and Body Image: Lessons from Phantom Limbs, Capgras Syndrome and Pain Asymbolia," *Phil. Trans. Roy. Soc. Lond.* B353:1851–9 (1998). For clinical and experimental details, see Ramachandran and Rogers-Ramachandran, "Synaesthesia in Phantom Limbs" (1996).

42. P. Brugger et al., "Beyond Re-membering: Phantom Sensations of Congenitally Absent Limbs," *Proc. Nat. Acad. Sci. USA* 97:6167–72 (2000).

43. See § 12 and § 13 of *The Ethics*.

CHAPTER 4

1. Adapted from the case report of a sixty-eight-year-old woman suffering from stroke-related, transient alien hand syndrome. From D. H. Geschwind et al., "Alien Hand Syndrome: Interhemispheric Disconnection Due to Lesion in the Midbody of the Corpus Callosum," *Neurology* 45:802–808 (1995).

2. See K. Goldstein, "Zur Lehre der Motorischen Apraxie," *Jour. für Psychologie und Neurologie* 11:169–187 (1908); W. H. Sweet, "Seeping Intracranial Aneurysm Simulating Neoplasm," *Arch. Neurology & Psychiatry* 45:86–104 (1941); S. Brion & C.-P. Jedynak, "Troubles du Transfert Interhémisphérique (Callosal Disconnection). A Propos de Trois Observations de Tumeurs du Corps Calleux. Le Signe de la Main Étrangère," *Revue Neurologique* 126:257–266 (1972); G. Goldberg et al., "Medial Frontal Cortex Infarction and the Alien Hand Sign," *Arch. Neurology* 38:683–686 (1981). For an important new conceptual distinction, see C. Marchetti & S. Della Sala, "Disentangling the Alien and the Anarchic Hand," *Cog. Neuropsychiatry* 3:191–207 (1998).

3. Goldberg et al., "Medial Frontal Cortex Infarction," 684 (1981).

4. G. Banks et al., "The Alien Hand Syndrome: Clinical and Postmortem Findings," *Arch. Neurology* 46:456–459 (1989).

5. Ibid.

6. For more on the representational architecture of volition and akinetic mutism, see T. Metzinger, "Conscious Volition and Mental Representation: Towards a More Fine-Grained Analysis," in Natalie Sebanz & Wolfgang Prinz, eds., *Disorders of Volition* (Cambridge, MA: MIT Press, 2006).

7. S. Kremer et al., Letter to the Editor, "The Cingulate Hidden Hand," *Jour. Neurology, Neurosurgery, and Psychiatry* 70:264–265 (2001); see also a classical study by I. Fried et al., "Functional Organization of Human Supplementary Motor Cortex Studied by Electrical Stimulation," *Jour. Neurosci.* 11:3656–66 (1991). In this study, subjects stimulated with electrical currents of different strength reported the illusory conscious perception of ongoing movement, or the anticipation of movement, or the "urge" to perform a movement, all "in the absence of overt motor activity."

8. D. M. Wegner & T. Wheatley, "Apparent Mental Causation: Sources of the Experience of Will," *Amer. Psychol.* 54(7):480–492 (1999).

9. Wegner & Wheatley, "Apparent Mental Causation" (1999), 488.

10. Ibid., 483.

11. See, for instance, P. Haggard, "Conscious Awareness of Intention and of Action," in Johannes Rössler & Naomi Eilan, eds., *Agency and Self-*

Awareness—Issues in Philosophy and Psychology (Oxford, UK: Clarendon Press, 2003). A good recent review is Patrick Haggard, "Human Volition: Towards a Neuroscience of Will," *Nat. Rev. Neurosci.* 9:934–946 (2008).

12. It is true that indeterminacy exists on the subatomic level, but the mind cannot somehow sneak into the physical world through indeterminate quantum processes. (Nor is chance what we want: Philosophically, randomness in the brain would be just as bad as full determination.) Quantum theories of free will are empirically false as well: There may be different kinds of brains somewhere else in the universe, but in human brains the firing of neurons and so on take place on the macroscopic scale. For such huge objects as nerve cells at 37°C body temperature, quantum events simply play no role.

13. The voluntary inhibition of voluntary actions seems to be mostly determined by unconscious events in the anterior median cortex. See M. Brass & P. Haggard, "To Do or Not To Do: The Neural Signature of Self-Control," *J. Neurosci.* 27:9141–9145. (2007).

14. See T. Metzinger, "The Forbidden Fruit Intuition," The *Edge* Annual Question—2006: What Is Your Dangerous Idea? www.edge.org/q2006/q06_7.html#metzinger. Reprinted in J. Brockman, ed., *What Is Your Dangerous Idea? Todays's Leading Thinkers on the Unthinkable* (New York: HarperPerennial, 2007), 153–155.

15. It would not be a new thought in the history of philosophy. Vasubandhu, a fourth-century Buddhist teacher and one of the most important figures in the development of Mahayana Buddhism in India, reports: *Buddha has spoken thus: 'O, Brethren! actions do exist, and also their consequences (merit and demerit), but the person that acts does not. There is no one to cast away this set of elements and no one to assume a new set of them. (There exists no individual), it is only a conventional name given to (a set) of elements.'* Appendix to the VIIIth chapter of Vasubandhu's Abhidarmakoça, §9: 100.b.7; quoted after T. Stcherbatsky, "The Soul Theory of the Buddhists," *Bull. Acad. Sci. Russ.* 845 (1919).

CHAPTER 5

1. The second question, of course, is the one Descartes asked in the first Meditation, when he realized that everything he had ever believed to be certain—including his impression of sitting by the fire in his winter coat and closely inspecting the piece of paper in his hands—could equally well have

occurred in a dream. What makes the problem of dream skepticism so intractable is that even in a "best-case scenario" of sensory perception, there is apparently no reliable, fool-proof method of distinguishing wakefulness and dreaming. According to dream skepticism, literally all of our experiences of waking life could be nothing more than a dream, and we are unable, even in principle, ever to decide this question with certainty. For a detailed discussion of the problem of dream skepticism, see, for instance, Barry Stroud, *The Significance of Philosophical Scepticism* (New York: Oxford University Press, 1984). For the status of the phenomenal and the epistemic subject in the dream state, see J. Windt & T. Metzinger, "The Philosophy of Dreaming and Self-Consciousness: What Happens to the Experiential Subject During the Dream State?" in Patrick McNamara & Deirdre Barrett, eds., *The New Science of Dreaming* (Westport, CT: Praeger, 2007). See http://eprints.assc.caltech.edu/200/01/Dreams.pdf.

2. See J. A. Hobson et al., "Dreaming and the Brain: Toward a Cognitive Neuroscience of Conscious States," *Behavioral and Brain Sci.* 23:793–842 (2000); and Antti Revonsuo, *Inner Presence: Consciousness as a Biological Phenomenon* (Cambridge, MA: MIT Press, 2006).

3. Helen Keller, *The World I Live In* (New York: New York Review Books, 2003).

4. H. Bertolo et al., "Visual Dream Content, Graphical Representation and EEG Alpha Activity in Congenitally Blind Subjects," *Cog. Brain Res.* 15:277–284 (2003).

5. See C. H. Schenck, *"Violent Moving Nightmares,"* www.parasomnias-rbd.com/; E. J. Olson et al., "Rapid Eye Movement Sleep Behaviour Disorder: Demographic, Clinical, and Laboratory Findings in 93 Cases," *Brain* 123:331–339 (2000); and C. H. Adler & M. J. Thorpy, "Sleep Issues in Parkinson's Disease," *Neurology* 64 (suppl. 3):12–20 (2005).

6. See Hobson et al., "Dreaming and the Brain" (2000) for details.

7. F. van Eeden, "A Study of Dreams," *Proc. Soc. Psychical Res.* 26:431–461 (1913).

8. Oliver Fox, *Astral Projection* (New Hyde Park, NY: University Books, 1962). Also quoted in S. LaBerge & J. Gackenbach, "Lucid Dreaming," in Etzel Cardeña et al., eds., *Varieties of Anomalous Experience: Examining the Scientific Evidence* (Washington, DC: American Psychological Association, 2000).

9. See Paul Tholey, *Schöpferisch träumen* (Niedernhausen, Ger.: Falken Verlag, 1987).

10. See Stephen LaBerge & Howard Rheingold, *Exploring the World of Lucid Dreaming* (New York: Ballantine, 1990).

11. S. LaBerge et al., "Lucid Dreaming Verified by Volitional Communication During REM Sleep," *Perceptual and Motor Skills* 52:727–732 (1981); and S. LaBerge et al., "Psychophysiological Correlates of the Initiation of Lucid Dreaming," *Sleep Res.* 10:149 (1981).

12. For details, see P. Garfield, "Psychological Concomitants of the Lucid Dream State," *Sleep Res.* 4:183 (1975); S. LaBerge, "Induction of Lucid Dreams," *Sleep Res.* 9:138 (1980); S. LaBerge, "Lucid Dreaming as a Learnable Skill: A Case Study," *Perceptual and Motor Skills* 51:1039–41 (1980); LaBerge & Rheingold, *Exploring the World of Lucid Dreaming* (1990); and G. S. Sparrow, "Effects of Meditation on Dreams," *Sundance Comm. Dream Jour.* 1:48–49 (1976).

13. Hobson et al., "Dreaming and the Brain" (2000), 837. For details on the relation between the DLPFC and reflective thought, see A. Muzur et al., "The Prefrontal Cortex in Sleep," *Trends Cog. Sci.* 6:475–481 (2002).

14. Tholey, *Schöpferisch träumen* (1987), 97. English translation by T. Metzinger.

15. "Spandrels" refers to Stephen Jay Gould and Richard C. Lewontin's 1979 essay "The Spandrels of San Marco and the Panglossian Paradigm," in which, using the architectural analogy, the authors argue that some biological features are exaptations: that is, currently used for something other than what they were "developed for" during natural selection. *Proc. Royal Soc. London, Ser. B, Biol. Sci. (1934–1990)* 205(1161):581–598 (September 21, 1979).

CHAPTER 6

1. For details, see G. Rizzolatti et al., "From Mirror Neurons to Imitation: Facts and Speculations," in Andrew N. Meltzoff & Wolfgang Prinz, eds., *The Imitative Mind: Development, Evolution, and Brain Bases* (Cambridge: Cambridge University Press, 2002); and G. Rizzolatti & M. Gentilucci, "Motor and Visual-Motor Functions of the Premotor Cortex," in Pasko Rakic & Wolf Singer, eds., *Neurobiology of Neocortex* (New York: John Wiley & Sons, 1988). An excellent recent overview is Giacomo Rizzolatti & Corrado Sinigaglia, *Mirrors in the Brain: How Our Minds Share Actions and Emotions* (Oxford: Oxford University Press, 2008).

2. The mirror-neuron system may occasionally go awry. Patients suffering from a rare but well-known neurological syndrome called *echopraxia* are inevitably forced to act out any behavior they observe in other human beings. What likely happens in these patients is that the mirror-neuron system is inadvertently coupled to the motor system because of lack of prefrontal inhibition. Your mirror neurons come online and lose their ordinary status as purely offline simulators. Therefore, you are literally driven by the actions you see other people performing.

3. See V. Gallese & A. Goldman, "Mirror Neurons and the Simulation Theory of Mind-Reading," *Trends Cog. Sci.* 2:493–501 (1998); M. Iacoboni et al., "Cortical Mechanisms of Imitation," *Science* 268:2526–8 (1999); and V. Gallese, "The 'Shared Manifold' Hypothesis: From Mirror Neurons to Empathy," *Jour. Consciousness Studies* 8:33–50 (2001).

4. See T. Metzinger & V. Gallese, "The Emergence of a Shared Action Ontology: Building Blocks for a Theory," in G. Knoblich et al., eds., *Self and Action.* Special issue of *Consciousness & Cognition* 12(4):549–571 (2003).

5. V. Gallese, "Intentional Attunement: A Neurophysiological Perspective on Social Cognition and Its Disruption in Autism," *Brain Res.* 1079:15–24 (2006); F. de Vignemont & T. Singer, "The Empathic Brain: How, When, and Why?" *Trends Cog. Sci.* 10:435–441 (2006).

6. L. Carr et al., "Neural Mechanisms of Empathy in Humans: A Relay from Neural Systems for Imitation to Limbic Areas," *Proc. Nat. Acad. Sci. USA* 100(9):5497–5502 (2003); see also A. Goldman & C. S. Sripada, "Simulationist Models of Face-Based Emotion Recognition," *Cognition* 94:193–213 (2005).

7. A. D. Lawrence et al., "Selective Disruption of the Recognition of Facial Expressions of Anger," *NeuroReport* 13(6):881–884 (2002).

8. I. Morrison et al., "Vicarious Responses to Pain in Anterior Cingulate Cortex: Is Empathy a Multisensory Issue?" *Cog. Affec. & Behav. Neuroscience* 4:270–278 (2004); P. L. Jackson et al., "How Do We Perceive the Pain of Others: A Window into the Neural Processes Involved in Empathy," *NeuroImage* 24:771–779 (2005); M. Botvinick et al., "Viewing Facial Expressions of Pain Engages Cortical Areas Involved in the Direct Experience of Pain," *NeuroImage* 25:315–319 (2005).

9. This was the step from what I call *second-order embodiment* to *third-order embodiment.* In order to counteract the semantic inflation of the term "embodiment," I have introduced the notions of "first-order embodiment" (the bottom-up self-organization of intelligent behavior avoiding explicit

computation and relying only on physical properties of the system), "second-order embodiment" (generating intelligent behavior by using an integrated representation of the body as a whole, by internally representing oneself as embodied), and "third-order embodiment" (the functional elevation of second-order embodiment to the level of global availability, i.e., the *conscious experience* of embodiment). A short summary can be found in *Scholarpedia* 2 (10):4174 (2007) at www.scholarpedia.org/article/Self_Models.

10. V. Gallese, "Embodied Simulation: From Neurons to Phenomenal Experience," *Phen. Cog. Sci.* 4:23–38 (2005).

11. Gallese calls this specific phenomenal state "intentional attunement"—the peculiar experiential quality of familiarity with other individuals that arises because we implicitly match their intentions with processes that go on in our own brain when we form such intentions.

12. See T. Metzinger, "Self Models," *Scholarpedia* 2(10):4174 (2007) at www.scholarpedia.org/article/Self_Models; and Metzinger, "Empirical Perspectives from the Self-Model Theory of Subjectivity," *Progress in Brain Res.* 168:215–246 (2008).

13. See W. B. Carpenter, *Principles of Mental Physiology* (London: Routledge, 1875). For a review, see H. Richter, "Zum Problem der ideomotorischen Phänomene," *Zeit. für Psychologie* 71:161–254 (1957).

14. T. Lipps, "Einfühlung, innere Nachahmung und Organempfindung," *Arch. der Psychologie* 1:185–204 (1903).

15. See G. Rizzolatti & Laila Craighero, "The Mirror-Neuron System," *Ann. Rev. Neurosci.* 27:169–192 (2004); the classical paper is Rizzolatti & M. A. Arbib, "Language Within Our Grasp," *Trends Neurosci.* 21:188–194 (1998). For a brief first overview, see Rizzolatti & Destro, "Mirror Neurons," *Scholarpedia* 3(1):2055 (2008).

16. See Rizzolatti & Destro, "Mirror Neurons"; www.scholarpedia.org/artical/Mirror_neurons.

17. See Gallese, "The 'Shared Manifold' Hypothesis" (2001), for an additional discussion, see pp. 174 of this book.

18. Jerome S. Bruner, *Acts of Meaning* (Cambridge, MA: Harvard University Press, 1990), 40.

CHAPTER 7

1. http://technology.newscientist.com/article.ns?id=mg19926696.100&print=true.

2. A. Cleeremans, "Computational Correlates of Consciousness," *Prog. Brain Res.* 150:81–98 (2005). See also his "Consciousness: The Radical Plasticity Thesis," *Prog. Brain Res.* 168:19–33 (2008).

3. J. Bongard et al., "Resilient Machines Through Continuous Self-Modeling," *Science* 314:1118–21 (2006).

4. Ibid. In particular, see also free online support material at www.sciencemag.org/cgi/content/full/314/5802/1118/DC1. (See www.ccsl .mae.cornell.edu/research/selfmodels/morepictures.htm for additional online material.)

5. See also Thomas Metzinger, "Empirical Perspectives from the Self-Model Theory of Subjectivity: A Brief Summary with Examples," in Rahul Banerjee & Bikas K. Chakrabarti, eds., *Progress in Brain Research* (Amsterdam: Elsevier, 2008) 168:215–246. DOI: 10.1016/S0079–6123(07)68018–2.

6. Karl Popper & J. C. Eccles, *The Self and Its Brain: An Argument for Interactionism* (New York: Routledge, 1984), 208. Alan M. Turing's paper is in *Mind* 59:433–460 (1950).

7. It is interesting to note how perhaps the foremost theoretical "blind spot" of current philosophy of mind is conscious suffering. Thousands of pages have been written about color qualia and zombies, but almost no theoretical work is devoted to ubiquitous phenomenal states such as physical pain, boredom, or the everyday sadness known as subclinical depression. The same is true of panic, despair, shame, the conscious experience of mortality, and the phenomenology of losing one's dignity. Why are these forms of conscious content generally ignored by the best of today's philosophers of mind? Is it simple careerism ("Nobody wants to read too much about suffering, no matter how insightful and important the arguments are"), or are there deeper, evolutionary reasons for this cognitive scotoma? When one examines the ongoing phenomenology of biological systems on our planet, the varieties of conscious suffering are at least as dominant as, say, the phenomenology of color vision or the capacity for conscious thought. The ability to consciously see color appeared only very recently, and the ability to consciously think abstract thoughts of a complex and ordered form arose only with the advent of human beings. Pain, panic, jealousy, despair, and the fear of dying, however, appeared millions of years earlier and in a much greater number of species.

CHAPTER 8

1. Our belief in invisible persons may have different roots, possibly including so-called hyperactive agent-detection devices (see D. Barrett, "Exploring the Natural Foundations of Religion," *Trends Cog. Sci.* 4:29–34, 2000) and ancestor cults: See Daniel C. Dennett, *Breaking the Spell: Religion as a Natural Phenomenon* (New York: Viking, 2006), esp. 109ff; and Thomas Metzinger, *Being No One: The Self-Model Theory of Subjectivity* (Cambridge, MA: MIT Press, 2003), 371ff. Also note that out-of-body experiences would almost inevitably have contributed to early humankind's firm belief in the existence of invisible persons and more subtle levels of reality. See T. Metzinger, "Out-of-Body Experiences as the Origin of the Concept of a 'Soul,'" *Mind and Matter* 3(1):57–84 (2005).

CHAPTER 9

1. Y. Kamitami & S. Shimojo, "Manifestation of Scotomas Created by Transcranial Magnetic Stimulation of Human Visual Cortex," *Nature Neuroscience* 2:767–771 (1999).

2. B.-P. Bejjani et al., "Transient Acute Depression Induced by High-Frequency Deep-Brain Stimulation," *N.E. Jour. Med.* 340:1476–80 (1999). Here are examples of how the patient described her own conscious experience: "I'm falling down in my head, I no longer wish to live, to see anything, hear anything, feel anything." The authors report that when she was asked why she was crying and if she felt pain, she responded: "No, I'm fed up with life, I've had enough. . . . I don't want to live anymore, I'm disgusted with life. . . . Everything is useless, always feeling worthless, I'm scared in this world." When asked why she was sad, she replied: "I'm tired. I want to hide in a corner. . . . I'm crying over myself, of course. . . . I'm hopeless, why am I bothering you." Note that deep brain stimulation can also have just the opposite effect, namely, relief from serious, treatment-resistant depression. Here is a description: "All patients spontaneously reported acute effects including 'sudden calmness or lightness,' 'disappearance of the void,' 'sense of heightened awareness,' 'connectedness,' and sudden brightening of the room, including a sharpening of visual details and intensification of colors in response to stimulation." See H. Mayberg, "Clinical Study: Deep Brain Stimulation for Treatment-Resistant Depression," *Neuron* 45:651–660 (2005).

3. C. M. Cook & M. A. Persinger, "Experimental Induction of the 'Sensed Presence' in Normal Subjects and an Exceptional Subject," *Percept. Mot. Skills* 85:683–693 (1997). For a critical assessment and self-experiential report, see John Horgan, *Rational Mysticism: Dispatches from the Border Between Science and Spirituality* (New York: Houghton Mifflin, 2003).

4. See M. A. Persinger, "Religious and Mystical Experiences as Artifacts of Temporal Lobe Function: A General Hypothesis," *Perc. Mot. Skills* 57:1255–62 (1983). Clinicians have long observed a deepening of emotionality plus the development of a serious, highly ethical and spiritual demeanor in certain patients with chronic mesial temporal lobe epilepsy. Whether this can count as evidence for a specific kind of "personality syndrome" is still disputed. See O. Devinsky & S. Najjar, "Evidence Against the Existence of a Temporal Lobe Epilepsy Personality Syndrome," *Neurology* 53:S13–S25 (1999); D. Blume, "Evidence Supporting the Temporal Lobe Epilepsy Personality Syndrome," *Neurology* 53:S9–S12 (1999).

5. For more information, see the Web portal my collaborators Carsten Griesel and Elisabeth Hildt have created at www.neuroethics.uni-mainz.de. For a recent overview, see T. Metzinger & E. Hildt, "Cognitive Enhancement," Judy Illes, ed., *Oxford Handbook of the Neuroethics* (Oxford: Oxford University Press 2010).

6. B. Maher, "Poll Results: Look Who's Doping," *Nature* 452:674–675 (2008). See also B. Sakhanian & S. Morein-Zamir, "Professor's Little Helper," *Nature* 450:1157–5 (2007).

7. M. J. Farah et al., "Neurocognitive Enhancement: What Can We Do and What Should We Do?" *Nature Reviews Neuroscience* 5:421–425 (2004). Four years later, after a careful analysis of pros and cons and perhaps surprisingly to many, leading figures in neuroethics are now coming to the conclusion that "We should welcome new methods of improving our brain function. In a world in which human workspans and lifespans are increasing, cognitive enhancement tools—including the pharmacological—will be increasingly useful for improved quality of life and extended work productivity, as well as to stave off normal and pathological accerelated cognitive declines. Safe and effective cognitive enhancers will benefit both the individual and society." See H. Greely et al., "Towards Responsible Use of Cognitive-Enhancing Drugs by the Healthy," *Nature* 456:702–705 (2008).

8. For a first and careful discussion of this important question, see T. Douglas, "Moral Enhancement," *J. Appl. Phil.* 25:228–245 (2008).

9. Alexander Shulgin and Ann Shulgin, *PiHKAL: A Chemical Love Story* (Transform Press, 1991); and Alexander Shulgin, *TiHKAL: The Continuation* (Transform Press, 1997).

10. W. N. Pahnke & W. A. Richards, "Implications of LSD and Experimental Mysticism," *Jour. Religion & Health* 5:179 (1966).

11. R. R. Griffiths et al., "Psilocybin Can Occasion Mystical-Type Experiences Having Substantial and Sustained Personal Meaning and Spiritual Significance," *Psychopharm.* 187:268–283 (2006).

12. R. R. Griffiths et al., "Mystical-Type Experiences Occasioned by Psilocybin Mediate the Attribution of Personal Meaning and Spiritual Significance 14 Months Later," *Jour. Psychopharm.* 22:621–632 (2008).

13. N. Malleson, "Acute Adverse Reactions to LSD in Clinical and Experimental Use in the United Kingdom," *Br. Jour. Psychiatry* 118:229–230 (1971).

14. S. Cohen, "Lysergic Acid Diethylamide: Side Effects and Complications," *Jour. Nerv. Ment. Dis.* 130:30–40 (1960).

15. See R. J. Strassman, "Adverse Reactions to Psychedelic Drugs: A Review of the Literature," *Jour. Nerv. Men. Dis.* 172:577–595 (1984); J. H. Halpern & H. G. Pope, "Do Hallucinogens Cause Residual Neuropsychological Toxicity?" *Drug Alcohol Depend.* 53:247–256 (1999); M. W. Johnson et al., "Human Hallucinogen Research: Guidelines for Safety," *Jour. Psychopharm.* 22:603–620 (2008). In the most recent and comprehensive review of the scientific literature, the authors actually make the interesting (and perhaps bold) claim that "The incidence of psychotic reactions, suicide attempts, and suicides during treatment with LSD [...] appears comparable to the rate of complications during conventional psychotherapy." See Torsten Passie et al., "The Pharmacology of Lysergic Acid Diethylamide: A Review," *CNS Neuroscience & Therapeutics* 14:295–314 (2008).

16. B. Vastag, "Poised to Challenge Need for Sleep, 'Wakefulness Enhancer' Rouses Concerns," *Jour. Amer. Medic. Assoc.* 291(2):167 (2004).

17. See Judy Illes, *Neuroethics: Defining the Issues in Theory, Practice, and Policy* (New York: Oxford University Press, 2005); and P. R. Wolpe et al., "Emerging Neurotechnologies for Lie-Detection: Promises and Perils," *Amer. Jour. Bioethics* 5(2):39–49 (2005); or T. Metzinger, "Exposing Lies," *Scientific American MIND,* October/November:32–37 (2006).

18. Haim Harari, "Democracy May Be on Its Way Out" (2006), www.edge.org/q2006/q06_2.html#harari.

INDEX